T0258130

Fiber Optic Sensors

Fiber Optic Sensors

Edited by **Bob Tucker**

New York

Published by NY Research Press,
23 West, 55th Street, Suite 816,
New York, NY 10019, USA
www.nyresearchpress.com

Fiber Optic Sensors
Edited by Bob Tucker

International Standard Book Number: 978-1-63238-194-1 (Hardback)

Printed in the United States of America.

Contents

Preface

This book was inspired by the evolution of our times; to answer the curiosity of inquisitive minds. Many developments have occurred across the globe in the recent past which has transformed the progress in the field.

All recent discoveries and explorations related to fiber optical sensors technology in the recent past have been thoroughly discussed in this book. This book provides a detailed study about latest developments of the subject, essential propositions of varied sensor types, their role in structural health monitoring and evolution of varied physical, chemical and biological criteria. By presenting various advanced styles of sensing and systems, and by depicting latest evolvements in fiber Bragg-grating, reflectometry and interfometry based sensors; the book highlights the improvement and growth of fiber optics sensors. This book successfully unites academicians and practitioners in an overall observation of the subject. This book can play a pivotal role by providing the required information to scientists working and teaching, optical fiber sensor technology' and it can also be used by industrialists who need to update themselves with modern and recent progresses in the field.

This book was developed from a mere concept to drafts to chapters and finally compiled together as a complete text to benefit the readers across all nations. To ensure the quality of the content we instilled two significant steps in our procedure. The first was to appoint an editorial team that would verify the data and statistics provided in the book and also select the most appropriate and valuable contributions from the plentiful contributions we received from authors worldwide. The next step was to appoint an expert of the topic as the Editor-in-Chief, who would head the project and finally make the necessary amendments and modifications to make the text reader-friendly. I was then commissioned to examine all the material to present the topics in the most comprehensible and productive format.

I would like to take this opportunity to thank all the contributing authors who were supportive enough to contribute their time and knowledge to this project. I also wish to convey my regards to my family who have been extremely supportive during the entire project.

Editor

Long Period Fiber Grating Produced by Arc Discharges

Julián M. Estudillo-Ayala[1], Ruth I. Mata-Chávez[1],
Juan C. Hernández-García[2] and Roberto Rojas-Laguna[1]
[1]*Universidad de Guanajuato*
[2]*Centro de Investigaciones en Óptica A.C.*
México

1. Introduction

Fiber Gratings are found among the most popular devices widely used in both, optical communications and optical fiber sensing. Within optical communication networks, optical fiber-based devices perform critical operations such as coupling/splitting, wavelength-selective filtering, and optical switching (Han, Y. G. et al., 2004; An H. et al., 2004; Eggleton B. J. et al., 1997). In the field of optical sensing, optical fiber sensors, with unique advantages such as immunity to electromagnetic interference, high sensitivity, resistance to corrosion, and high temperature survivability, are widely used to measure various physical variables such as stress, temperature, pressure, refractive index, etc. and chemical parameters (Udd, E. 1991).

There are two types of fiber gratings that have been developed so far including the Fiber Bragg Grating (FBG) and the Long Period Fiber Grating (LPFG). One important advantage respect to Bragg gratings is that their fabrication process is simpler; their cost is lower. Furthermore, they present low retro-reflection and high sensitivity in sensing applications (Kersey, A. D. et al., 1997).

LPFGs have been traditionally fabricated by exposing photosensitive optical fiber to ultraviolet (UV) light transversely. Either through an amplitude mask or point-by-point to create a periodic refractive index change inside an optical fiber. Unfortunately, the UV-induced refractive index changes inside an optical fiber can only survive at a relatively low temperature. As a result, the UV exposure fabricated-LPFGs are not suitable for high temperature applications. Among the different techniques available for the fabrication of LPFGs, the electric arc technique is one of the few enabling their fabrication, virtually in any type of glass fibers. This technique is simple, inexpensive and harmless when compared to the ones based on laser radiation. Furthermore, arc-induced gratings are suitable for high temperature applications since their formation mechanisms rely on thermal effects. Thus, the temperature reached by the fiber during an electric arc discharge, is a key parameter to understand the properties of gratings (Rego, G. 2010). An LPFG can be considered a special case of a fiber Bragg grating, in which the period of the index modulation satisfies a phase matching condition between the fundamental core mode and the forward-propagating cladding-mode of an optical fiber. Therefore, an LPFG consists of a periodic spatial variation (along the fiber longitudinal axis) in the refractive index of an optical fiber. The periodic refractive index modulation couples light from a forward-propagating core-guided mode, to forward-propagating cladding-guide modes, near certain resonance wavelengths (Vengsarkar, A. M. et al., 1996). The light coupled into the cladding modes, eventually attenuates due to the high loss of the cladding modes. As a result, the transmission spectrum of a LPFG has a series of discrete attenuation bands near the resonance wavelengths. In this chapter we will describe two of the in-fiber LPFGs fabrication methods: optical fiber fattening and optical fiber micro-taper, both methods by means of electric arc (we used a splicer fusion of optical fiber). We will present several configurations of PFGs sensors as well as temperature and curvature sensors. Finally, at the end of this chapter, we will show numerical calculations of the modal behavior of the LPFGs, besides conclusions and references.

2. Fiber grating

One of the most important discoveries, in the field of optical fibers, has undoubtedly been made by Hill et al. in 1978 (Hill, K.O. et al., 1993; Hill, K.O. et al., 1997). Working in their laboratories [The Communications Research Center (CRC) in Ottawa, Canada.], they carried out an experiment to study nonlinear effects in an optical fiber especially designed by Koichi Abe of Bell Northern Research. This special optical fiber had a high germanium concentration, a numerical aperture of 0.22 at 514.5 nm, and the fiber core diameter was of 2.5 mm. They launched intense light of Argon-ion (514.5 nm) laser into the core of a Germania-doped fiber, and several minutes after, they observed the power beam decreasing. Subsequent to this observation, they monitored the light back reflected from the fiber, finding that the power significantly increased with time. The increase in the reflected light was produced by the writing in the core of the fiber by an interference periodic pattern (hologram), corresponding to a periodic modulation of the refraction index in the optical fiber core. This nonlinear effect in optical fiber was called fiber photosensitivity, and the holograms in the core of the fiber are called "Hill gratings" (Hill & Meltz, 1997; Othonos, 1997; Hill, 2000).

After the discovery of the photosensitivity in optical fibers, the effect was forgotten by several years, due to restrictions in the writing of the grating. However, a new interest emerged after the proposal of Meltz et al. 1989, where they described the grating fabrication, producing the overlapping of two coherent beams of ultraviolent light (Meltz, G., et al., 1989). Beams interfered between them, creating a periodic interference pattern with both bright and dark bands, which generated a permanent grating with the same refraction index of the fiber. This technique was called the *transverse holographic technique*. This is possible because of the cladding of the fiber is transparent to the ultraviolet radiation, while the core strongly absorbs this kind of light due to the germanium concentration. Hill et al. 1993 proposed a simple way to fabricate gratings. They did not use an interferometer; instead, they used a beam of ultraviolet light placing a glass phase mask of pure silica, very close to the fiber (Othonos, A. et al. 1997). The grating was formed by exposing the fiber to ultraviolet light through the phase mask. Therefore, it is one of the most efficient methods for inscribing gratings in photosensitive fiber.

This novel method represented the key for simpler and more accelerated fabrication of gratings in special fibers. Nowadays, these gratings are employed in the fabrication of devices for optical communications (fiber lasers, filters, optical switching, and multiplexers) and optical sensing applications.

2.1 Basic concepts

The optical fiber gratings are a special type of filter wavelength selective operating in reflection or transmission with the fundamental principles of reflection, refraction and diffraction of light in periodic structures dimensionless. These basic components are formed when the refractive core index of a fiber section is periodically modulated. According to their operation, there are two types of optical fiber grating: Bragg Gratings (BG) and Long-Period Gratings (LPG). The Bragg gratings are called this way because their operation is based on the Bragg reflection law (Kashyap R., 1999). The Bragg gratings were inscribed by the first time in fiber optics, with the discovery of photosensitivity as it was mentioned previously. The main characteristic of this type of BG is the possibility to reject a narrow band of wavelengths (which meet the Bragg condition). Fig. 1 shows a diagram which illustrates the operation of a Bragg grating. When the signal intensity P_I (broadband light source) is propagated along the BG in the optical fiber, a signal P_B centered at the Bragg wavelength and a narrow band is reflected(< 1nm). On the other hand, the transmitted signal P_T, shows a band rejection to Bragg wavelength.

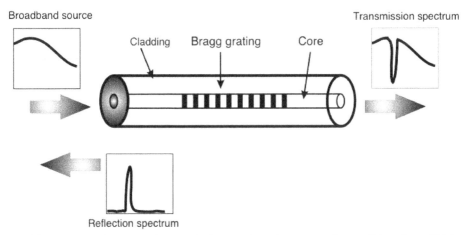

Fig. 1. (a) Schematic diagram of a sampled FBG. Signal spectra of input(P_I), reflected(P_B) and transmitted in a Bragg grating(P_T).

3. Long Period Fiber Grating

The long period fiber gratings (LPFG) has a period typically in the range from 100µm to 1000 µm(1mm), as we show in Fig. 2. An LPFG introduces a coupling between the fundamental guided mode and propagating cladding mode (Riant, I. 2002). In Fig. 3, we show the transmission spectrum of a LPFGs photoimprinted in a standard fiber. The wavelengths of the peaks shown in Fig. 3 are defined as (Erdogan, T. 1997).

$$\lambda^m = (n_{eff}^{01} - n_{eff}^m)\Lambda \tag{3.1}$$

Where n_{eff}^{01} and n_{eff}^{m} represent the effective indices of the fundamental guided mode, the cladding mode of order m coupled to the guide mode by the grating respectively, then Λ is the period of the LPFG.

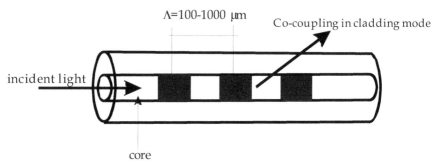

Fig. 2. Schematic diagram of a long-period fiber grating. The period of the grating is Λ.

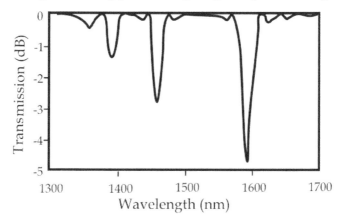

Fig. 3. Depicts transmission spectrum characteristic of a Long Period Grating

Coupling between two propagating modes occurs for periodicity of several hundreds of microns, which are three orders of magnitude larger than for short period grating.

The minimum transmission of the attenuation bands is governed by (Kashyap, R. 1999).

$$T_i = 1 - \sin^2(\kappa_i L) \tag{3.2}$$

Where L is the lenght of the LPFG and κ_i is the coupling coefficient for the ith cladding mode, which is determined by the integral overlap of the core and cladding mode, also by the amplitude of the periodic modulation of the mode propagation constants.

Since the cladding generally has a large radius, it supports a large number of cladding modes. Theoretical analysis has shown that efficient coupling is possible only between core and cladding modes which have a large integral overlap, in other words, modes which have similar electric field profiles (Erdogan, T. 1997). This coupling is observed between the core and circularly symmetric cladding modes of odd order. The effect occurs because the electric field profile of the even-order modes is such that the field amplitude is low within the core (Erdogan, T. 1997). The refractive index of the propagating core mode of the fiber is generally determined using the weakly guided field approximation (Glogle, D. 1971). A simple three-layer slab waveguide model, to determine the cladding mode indices, has been presented, allowing an approximate determination of the coupling wavelengths (Vengsarkar A.M. et al., 1996).

3.1 Fabrication techniques of long-period gratings in optical fiber

The long period grating fabrication in an optical fiber consists of a refractive index periodic modulation of the optical fiber core. This modulation may be permanent or temporary. The permanent modulation can be obtained by irradiation or heating. The temporal modulation index can be achieved by mechanical pressure or acoustic vibrations through the photo-elastic effect. The modulation of the core refractive index has been achieved by being used with different types of optical fibers, for example, standard single mode, photo-sensible, photonic crystal fiber, specialty optical fibers, among others. Techniques like UV irradiation (Ultraviolet) (Vengsarkar A.M. et. al., 1996; Bhatia, V. et. al. 1996; Erdogan, T. 1997), laser CO_2 exposure (Davis, D. D., et. al., 1998; Davis, D. D., et. al., 1999), electric arc discharge, mechanical microbends, etched corrugations, ion beam implantation, femtosecond laser exposure, just to mention a few, have been developed for more than a decade. Different devices based on LPFG have been fabricated for applications in optical communications as well as in sensors.

3.2 Techniques of fabrication by arc discharges

The Long Period Fiber Gratings (LPFG) fabricated with electric arc, have been widely used as filters in telecommunication networks, as well as the sensing element for various physical and chemical parameters. These applications are possible due to their relevant characteristics, such as: very deep attenuation band and narrow bandwidth. In addition, the period is easily adjustable, allowing the wavelength to tune over a wide range (from 1300 to 1700 nm). However, disadvantages of the LPFGs are their low repeatability and the wavelength of the attenuation bands has higher level of losses compared to other technologies (UV-inscribed techniques). Disadvantages are largely due to two issues: first, to the random nature of the arc, and second, the physical deformation that occurs in the optical fiber. In this chapter, we describe the development of LPFG when electric arc discharge is applied. The electric arc can produce geometric deformation, which might increase the coupling of modes, but in many cases it induces losses. There are three main types of modulation: modulation only of the refractive index, the modulation by microtaper, and microbends.

3.3 Modulation only of the refractive index

The best way to inscribe an LPFG is through merely changing the refractive index of the core, without affecting the geometry of the fiber. In this case, the attenuation bands are well defined and out of these ones, there are no losses. Fig. 4 shows the transmission spectrum of a long-period grating inscribed by ultraviolet radiation (LPFG-UV), this kind of gratings meet these conditions. In the LPFG by electric arc (LPFG-EA) this type of modulation can be obtained, but unlike LPFG-UV, in this type of gratings the refractive index of the cladding is

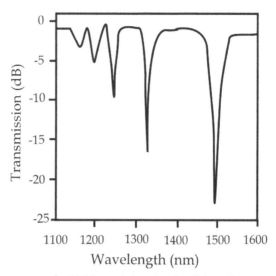

Fig. 4. Transmission spectra of a LPFG modulated only of the refractive index (LPFG-UV) (Stephen, W.J., et. al., 2003).

always modulated and geometric changes occur. The modulation of both refractive indices increases insertion losses, because the single mode condition is different, and moreover, the coupling constant is larger. The latter means that in general, the LPFG-EAs are of shorter length than LPFG-UV and easily, achieve greater levels of attenuation. While LPFG-UV typically require a grating length of 1 inch to achieve a rejection level of 18 dB, the LPFG-EA require approximately 2 cm.

The pure Refractive Index Modulation (RIM), in the electric arc technique, is quite difficult to achieve, as usually, it uses axial stress for greater change in refractive index, producing micro-bending or micro-tapers.

3.4 Periodic modulation by micro-taper

During an electrical arc discharge, an optical fiber can reach temperatures above 1350 ° C (Rego G., 2006). This temperature is sufficient to soften the silica (SiO_2) and make it pliable. This is the principle used for taper manufacture of optical fiber and fiber couplers couplers. This procedure to make a fiber taper is used to inscribe a LPFG. Due to the low weight (less than 200 g) and short electric discharge (hundreds of milliseconds), the tapered length is in the order of micrometers. So, actually, a micro-taper is achieved by every arc discharge, see fig. 5.

Fig. 5. Micrography of the induced tapers (Myoungwon K., et al., 2002)

In the micro-tapers of fiber, the best option is to reduce the fiber in symetrically shape, as shown in Fig. 6(a). Unfortunately due to factors such as: misalignment, accidental twist or tension, it is not exactly axial, and then an LPFG can be obtained as shown in Fig. 6(b). This mismatch of fiber alignment generates losses as the electromagnetic fields cannot be guided properly. In addition, birefringence is induced because the core is deformed asymmetrically. These features are present in micro-tapers. In the context of the LPFG-EAs, producing micro-tapers with a symmetrical shape is possible because the gratings allow good contrast with shorter length. Also, if handled properly, the procedure of micro-taper can be achieved with minimal levels of losses of RPLA similar to those obtained in other inscription techniques. The asymmetry can be reduced with a good alignment, while for narrow fiber, it is necessary to properly choose the parameters of the technique, in order to achieve an optimal micro-taper.

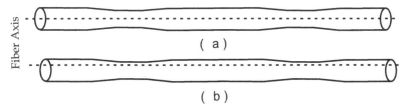

(a)

(b)

Fig. 6. Two types Microtapers of LPFG: (a) symmetric, (b) asymmetrical

3.5 Periodic modulation by microbends

The Modulation by microbends was presented by In Kag Hwang et al. (1999). In Fig. 7, an unjacketed optical fiber is held straight by two fiber holders, separated by 5.5 cm, which limits the maximum grating length. One of the fiber holders is then displaced (by 100 μm) in the orthogonal direction of the axis fiber so that a lateral stress is induced on the fiber between the two fiber holders. When a local section of the fiber is heated by application of an electric arc, the fiber is slightly deformed owing to the lateral stress, creating a microbend. The amplitude of the microbend is controlled by the duration of the arc and is typically less than 1 μm. Most of the originally applied lateral stress remains in the fiber.

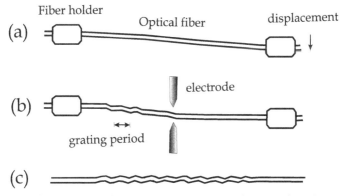

Fig. 7. Fabrication of a LPFG based on periodic microbends by use of an electric arc.(Hwang, I. K., et al. 1999).

Translating the electrodes by the grating period and applying electric arcs, one can create successive microbends without additional displacement of the fiber holder. In this way, a permanent microbend structure is inscribed in an optical fiber, as shown in Fig. 7(c).

Another way to induce microbends is applying many arc discharges at the same point and contrary to the microtapers, when applying the discharge; the glass is compressed axially over the fiber. This will result in a fattening of the area under the electric arc discharge. Fig. 8 shows a long period fiber grating, based in fiber fattening. These LPFGs   have very specific characteristics: the number of arc discharges are little, about 8 fattening points per grating and the attenuation bands reach a depth of almost 20 dB. At the same time, the length of the LPFG is very short (about 0.5 cm), but the level of losses that can be produce are at 1 to 2 dB (Mata-Chávez R.I., et al., 2008a).

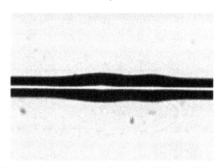

Fig. 8. Photograph of the fattened sections of the optical fiber. (Mata-Chávez R.I., et al., 2008).

3.6 Fabrication of LPFG by the micro-taper method

In these two sections we describe two fabrication methods of LPFGs by electric arc discharges: micro-tapers in which we show the experimental setup as well as data of the parameters used, and the method of fattening.

The gratings were fabricated in commercially avaible standard telecommunication optical fiber (SMF-28) using the point-by-point technique as shown in fig. 9. A bare single mode fiber without its protective coating was placed between the electrodes of a fusion splicer machine (Fitel S-175), while its ends were fixed on two Bare Fiber Adapters. A section of the fiber was fixed on a computer-controlled translation stage (Model NLS4, NEWMARK SYSTEMS INC.) wich a resolution of 1μm.

The fusion splicer is operated in manual mode, and parameters such as arc current, prefusion time, and pushing distance are varied in order to control the grating period. Best results were observed when in the splicer machine the arc current was set up to 11mA, the pre-fusion time was 150 ms, the arc duration time was of 250 ms and the Z-push was set to 1μs, which rise the temperature of the fiber section positioned in between the electrodes to its softening point. The fiber was then moved by the translation stage along a distance equal to the grating period and another arc discharge along with tapering was applied. Care was taken to provide sufficient time between arcs to allow the fiber to get cooled. The periodical tapering continued and the optical transmission of the fiber was monitored during the LPFG

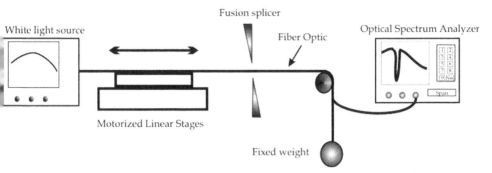

Fig. 9. Experimental setup used for fabrication LPFG by the micro-taper method.

fabrication process in order to obtain the desired spectral attenuation notches. The optical reading setup employed in the experiments were a white light source coupled to one end of the fiber, while the other end was connected to an optical spectrum analyzer, OSA (Yokogawa AQ6370B, Measurement wavelength range 600-1700nm), set to a resolution of 0.1 nm. The transmission spectrum was acquired and processed by a computer to determine the LPFG central wavelength of attenuation and transmittance. The size of the arc limited the minimum grating period Λ, and we got a peak attenuation of 22 dB at 1531 nm after 22 engravings with a grating period of 629μm. The spectral evolution of the LPFG is shown in Fig. 10. illustrating the transmission spectra with various grating period numbers of 10,15,17,20 and 21.

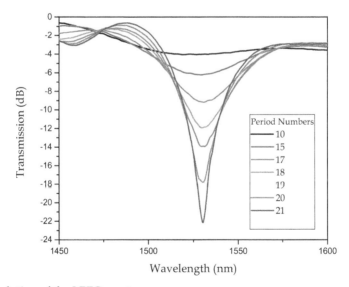

Fig. 10. Evolution of the LPFG spectrum

In fig. 11, it is shown spectra of LPFGs is depicted with different periods, we observed that for shorter periods, the attenuation peak shifts to the left, the LPFGs periods are Λ= 540nm, 560nm, 580nm and 600nm.

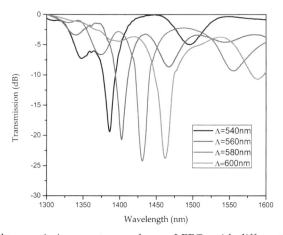

Fig. 11. Normalized transmission spectrum of some LPFGs with different periods.

3.7 Fabrication of LPFG by the fattening method

The fattening method used to fabricate long period fiber gratings consists in gradually enlarging the optical fiber diameter. This is achieved through applying several electrical discharges over an axial region and observing the transmission changes with an optical spectral analyzer. After applying the required discharges by the procedure in one point, the fiber is displaced by the motors of the fusion machine for a distance equal to the grating period (>150 microns) where the process is then repeated to enlarge again the fiber diameter (Mata-Chávez R.I., et. al., 2008a).

This procedure is repeated from two to six times (depending on the grating period) until filtering functions are observed. Fig. 12 depict the transmission loss band achieved by the method in a dispersion shifted fiber.

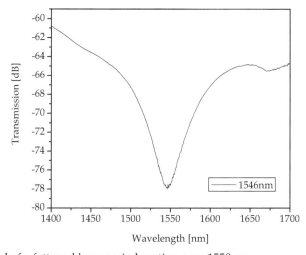

Fig. 12. Loss band of a fattened long period grating near 1550 nm.

The fusion splicer machine has resulted very useful for the fattening method. The standard fusion splice method is used to heat and soften the fiber (known as push delay) followed by a hot push to join the fiber tips (Yablon A. D. 2005). As the procedure is repeated the fiber gets a geometric deformation on its structure for which an energy coupling occurs of the fundamental mode of the fiber core to the radiation or leaky modes in the cladding. Fig. 13 depicts the time delay of the electric discharges applied to enlarge the fiber diameter in a point. Using this approach, wavelength band rejection filters have been obtained, this can be used as optical communication and sensing devices.

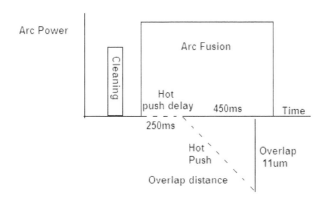

Fig. 13. Time delay of the standard discharged method.

The complete process is done in manual mode after choosing the correspondent optical type program and modifying several variables as arc power, arc duration, pre-fusion time and z-push distance. In fig. 13 can be observed an initial cleaning stage of low power to remove plastic and dust debris. Then, the fusion process continues with an electric arc discharge where the hot push delay heats the fiber to the softening point during 250ms. The hot push stage joins both tips by means of the motors of the machine for a distance of 11 microns. This process is repeated several times until a desired expansion of the diameter is achieved. The following values were initially used to fabricate LPFG in dispersion shifted fiber: Arc power = 85mW, Arc duration = 450ms, hot push delay = 250ms and z push = 11µm.

4. Devices produced by Long Period Fiber Grating

Several LPFG fabrication methods are used with different types of optical fibers like standard single mode, photo-sensible, photonic crystal fiber, specialty optical fibers, etc. Techniques like UV irradiation (Ultraviolet) (Vengsarkar A.M., et. al., 1996a; (Vengsarkar A.M., et al., 1996b; Erdogan, T., 1997), laser CO_2 exposure (Davis D. D., et. al., 1998; Davis D. D., et. al., 1999; Kakarantzas G., et. al., 2001), electric arc discharge (Humbert G., et al. , 2003; Dobb H., et. al., 2004; Rego G. M., et al., 2006b, mechanical microbends (Su L., et al., 2005), etched corrugations (Lin C. Y., et. al., 2001), ion beam implantation (von Bibra M. L., et. al., 2001), femtosecond laser exposure (Grobnic D., et al., 2006; Mihailov S. J., et al., 2006), etc. have been developed for more than a decade. Different devices based on LPFG have been fabricated for applications in optical communications and as sensors.

Sensing applications with LPFG include temperature, strain, bend, torsion, pressure and biochemical sensors because of the sensitivities of resonance wavelength and attenuation amplitude of these measured parameters. Variations of these measurands respond to the fabrication method and optical fiber type that influence on the grating length and optical characteristics of every sensor.

Optical telecommunications find useful applications for LPFG as band rejection filters (Mata-Chávez R.I., et. al., 2008a; Hernández-García J.C., et. al., 2010), gain equalizers for erbium doped fibers (Vengsarkar A.M. et al., 1996a; Vengsarkar A. M., et al., 1996b) , fiber dispersion compensators, core mode blockers, in fiber polarizers (Wang Y.P., et al., 2007), couplers (Liu Y., et. al, 2007), and mode converters (Hill K.O., et al., 1990), because of their bandstop filtering characteristics and selective mode coupling between the fundamental mode and the cladding modes.

5. Sensors of Long Period Fiber Grating

5.1 Curvature sensor

In this section the transmission effects of LPFGs by curves is shown, in Fig. 14 the experimental setup is depicted.

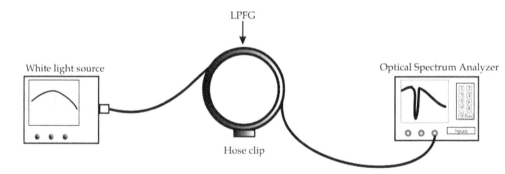

Fig. 14. Experimental setup for sensing diameter of curvature with LPFG by micro-taper modulation

For this experiment, it is necessary to maintain fixed the LPFGs and avoid twisting and LPFGs movements and optical fiber movements, these factors influence the behavior of the attenuation peak of LPFGs. We turned the screw and the hose clip diameter was varied, and then we were measuring the output spectrum of the LPFGs, measurements reduced the diameter of the loop from 13.1 to 12 cm. In Fig. 15, we show the displacement of the attenuation peak to the left as we decrease the diameter of the loop curvature.

In Fig.16, we show the almost linear behavior of the shift in the wavelength of the attenuation peak when the diameter of curvature of the LPFG was decreased. The attenuation peak is shifted by changing the period of the LPFGs. This is due to the change of the diameter of the hose clip. This scheme can serve as a curvature sensor and for tuning a fiber laser.

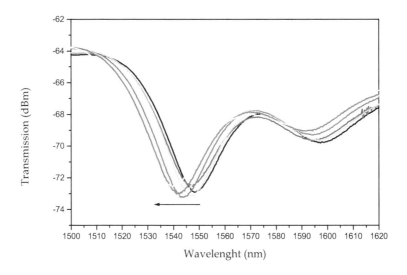

Fig. 15. displacement to left the attenuation peak when decrement the diameter of curvature.

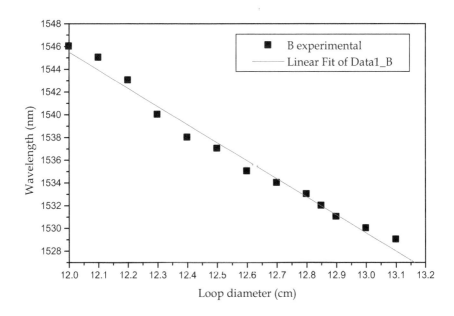

Fig. 16. Linear change of the attenuation peak

5.2 Temperature sensor

Long period gratings are inherently sensitive to changes in temperature of the surrounding environment. This can be observed when a full variation characterizations of the transmission spectra result in shifting of the attenuation band central wavelengths with a possible change in the peak intensity. The chain rule differentiation of the phase matching condition is a common approach to describe the temperature response of the LPFG (Mata-Chávez R.I., et al., 2008b).

$$\frac{d\lambda}{dT} = \frac{d\lambda}{d\left(\delta\left(n_{eff}\right)\right)} \left(\frac{dn_{eff}}{dT} - \frac{dn_{clad}}{dT} \right) + \Lambda \frac{d\lambda}{d\Lambda} \frac{1}{L} \frac{dL}{dT} \qquad (5.1)$$

In order to study the temperature effect over the transmission spectrum of the fattened LPG, the experimental setup shown in Fig. 17 was used. The characterization system consists on a fluorescence light source, which is obtained by pumping an Erbium Doped Fiber (EDF) with a laser diode (LD) at 980 nm. A temperature controller in a range of 50-500 °C, and an Optical Spectrum Analyzer (OSA) with a 0.2nm spectral resolution (connected with a dashed line in the drawing).

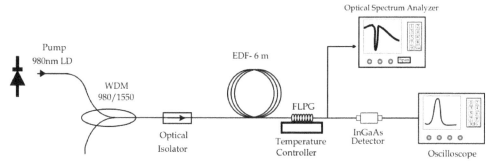

Fig. 17. Experimental setup for LPFG characterizations to study of temperature

When finishing the temperature characterizations, an optical detector and an oscilloscope were placed in the experimental setup, instead of the OSA, to measure the output in volts/100°C for the temperature sensor implementation. The detector is a cheaper and smaller device compared to the OSA that is coupled directly to the optical fiber. It is also a better option to read and process light signals with a smaller electronic acquisition data system.

The experimental setup with the OSA in the log scale shows a wavelength shift response which is linear as temperature increases. Resonance wavelength shifting was observed to longer wavelengths. This behavior is observed almost in all the studies presented in literature for temperature characterizations of LPFG in any type of optical fiber (Rego, G., et. al., 2005). The average sensitivity achieved is about 72 pm/°C. This is a greater value than the obtained in our previous work with DS-LPFG with a resonance wavelength around 1550nm where the fiber was fattened at positions separated by ~25μm (Mata-Chávez R.I., et al., 2008). We think the sensitivity increment is due to the wavelength set inside the loss band and the excited modes within that range as well as the change of the refractive effective index with temperature. As the structure changed, modal group dispersion is also modified, changing the thermo optic coefficients of the core and the ring, contributing to the sensitivity increment.

When switching to the linear scale (mW) , we observed two interesting situations which can be convenient to implement a temperature sensor around 1550 nm. The optical power of the light source was presented as a peak centered at 1560nm and 2.146mW. The LPFG with a resonant wavelength at 1524nm, spliced after the EDF, gradually attenuated the optical power as the loss band of the filter shifted to longer wavelengths with temperature increment. On the other hand, if the LPFG with a resonant wavelength at 1550nm was used, it instantly attenuated the power spectrum. As the loss band shifted to longer wavelengths, the power spectrum increased gradually with temperature (Fig. 18). This measurement approach gives two options to implement a temperature sensor. Attenuating or increasing the output optical power which depends on the position of the central wavelength of the attenuation band and the laser optical power spectrum. Linearity of the photodiode output voltage with temperature is observed in a range of 25-200°C. For temperature over 200-500°C, there is a visible increment of the dV/dT of the photodiode output. The highest sensitivity registered in volts is 0.05mV/°C and the results were repeatable. The insertion losses can be reduced with the improvement of the fabrication method of the DS-LPFG in future works.

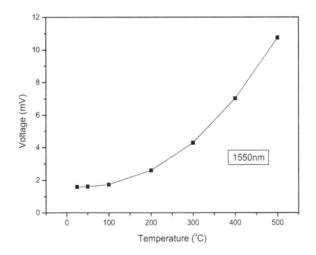

Fig. 18. Power spectrum increment using an LPFG with a central wavelength at 1550nm.

Then, it shows the characterization of an LPFG with micro-taper modulation, for temperature changes measurements. Once the LPFG was fabricated as described in section (3.6), we used a temperature controller (25 °C to 500 °C) for heating the grating. Light from a white light source was introduced to the LPFG.The spectrum for each measurement was obtained at the fiber end (50 °C, 100 °C-500 °C) as shown in the Fig. 19.The attenuation peak is shifted to the right when increasing the temperature from 25 °C to 500 °C. In Fig. 20, it is shown the linear behavior of the attenuation peak against temperature, we obtain those data for each Celsius degree, there was a wavelength variation of 63.15 pm .

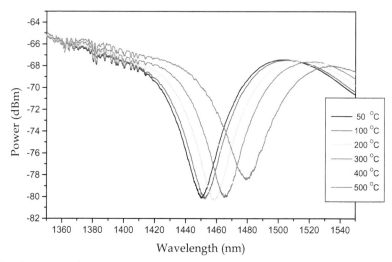

Wavelength (nm)

Fig. 19. Displacement of attenuation peak of LPFG to different temperatures.

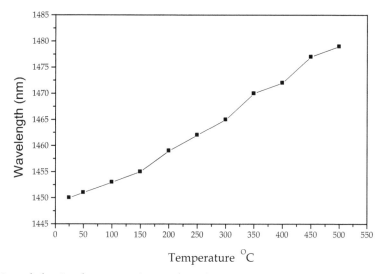

Temperature $^\circ$C

Fig. 20. Linear behavior the attenuation peak against temperature

6. Introduction to modelling individual Long Period Fiber Grating

There are several approaches for LPFG modeling to obtain the spectral properties. Some of them are based on previous studies of LPFG by Vengsarkar (1996), Erdogan, (1997), Kashyap, 1999, Othonos & Kalli, (1999). The coupled-mode theory has been widely applied to model LPFG where an excited mode by the grating can be represented as a linear combination of the eigenmodes of the fiber. The phase matching condition is also considered, where a perturbation of the phase of one mode would match the phase of another at the cladding region, see equation 3.1. A system of two coupled mode equations is given by:

$$\frac{dA_{co}(z)}{dz} = ik_{co-co}A_{co}(z) + i\frac{m_g}{2}k_{co-cl}A_{cl}(z)e^{i\delta c} \tag{6.1}$$

$$\frac{dA_{cl}(z)}{dz} = ik_{cl-co}A_{co}(z)e^{i2\delta c} + i\frac{m_g}{2}k_{cl.cl}A_{cl}(z) \tag{6.2}$$

A_{co} and A_{cl} are the slowly varying amplitudes of the core and cladding modes, k_{co-co}, k_{co-cl} are the coupling coefficients m_g the grating modulation depth and δ is the detuning from the resonant wavelength.

Numerical methods are also applied to the purpose of LPG analysis. It is the case of the Beam Propagation Method (BPM) and transfer-matrix method. These methods are also implemented in commercial software like BPM in RSoft®. With this software we have been able to model individual LPFG with fattening. Grating parameters and structure characteristics are introduced in the software to obtain the excited modes in the grating, the effective index, propagation through the grating, and power losses. A prime model considers an ideal complete structure for the LPFG in a dispersion shifted fiber Fig. 21. This fiber structure has a central core and a ring core which in a second part they are modeled by separate to obtain some characteristics as mode excitation. Different wavelengths were used which correspond to wavelength operation to common laboratory lasers. These are 1.55μm, 1.31μm, and 0.632μm as well as a broad band white light source (400-1700nm).

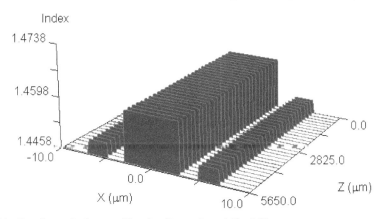

Fig. 21. Idealized step index profile of a dispersion shifted fiber.

The structure radii parameters were: $a = 2.9\mu m$, $b = 5\mu m$, $c = 8\mu m$ and the index refraction values : $n1=1.4598$, $n2=1.4498$, $n3=1.4498$ and $n4=1.4458$ (Mata-Chávez R.I., et al., 2008a). To investigate the complete modal characteristics, the fiber structure was divided in two regions as in reference (Mata-Chávez R.I., et. al., 2008b). A central core region and a ring core region, where light is confined and freely propagates along the structure. From this analysis it was encounter that only the LP_{01} and the LP_{11} modes are excited in the optical fiber. A model of the LPFG with fattening is also considered for the modal analysis and such structure is depicted in Fig. 22.

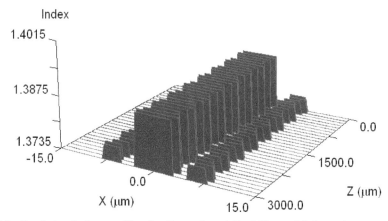

Fig. 22. Idealized step index profile of a dispersion shifted fiber with fattening.

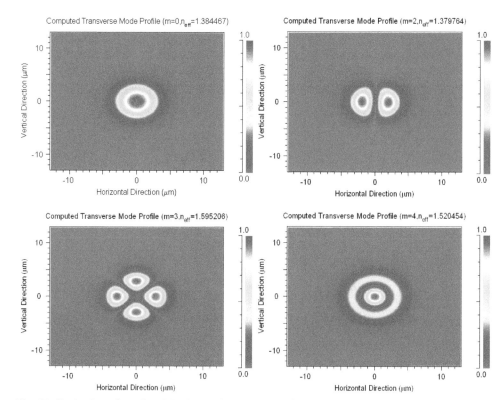

Fig. 23. Excited modes of an LPFG in a dispersion shifted fiber, LP_{01}, LP_{11}, LP_{21}, and LP_{03}.

Propagation of light can also be calculated. Fig. 24 depicts the LPFG with fattening. Several periods were modeled and power losses at the core can be achieved as light propagates. This is because of the energy that couples at the cladding layer due to the phase matching condition. The effective index of refraction can also be calculated as well as the LPFG transmission. This helps to compare the experimental results with the theoretical ones.

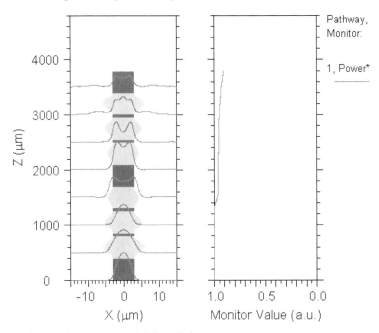

Fig. 24. Power loss at the core in a LPFG with fattening.

7. Conclusion

The main advantage of method of arc dischargers to fabricate long period fiber grating results from the fact that it can be implemented in any commercial fusion splicer, therefore, there are many potentialities of the electric arc technique to produce optical devices in an easy and fast way as sensors, filters and equalizers in any optical fiber. We presented the sensing of curvature and temperature, by means of long period fiber grating that we have fabricated by the method of micro-taper and fattening, and both present good sensitivity as much as 72pm/°C. This type of filter has potential applications in biochemical sensing, environment monitoring, stress, curvature and its principal advantages are that is compact, light and of low cost. One disadvantage of the electric arc method is the unrepeatability of the physical and optical characteristics of the fabricated filters due to the uncontrolled, in most cases, application of the electric arc discharges.

8. Acknowledgment

This work was supported by "División de Ingenierías del Campus Irapuato-Salamanca, Universidad de Guanajuato" and DAIP 2010 E20391.

9. References

An H., Ashton B., and Fleming S., (2004), Long-period-grating-assisted optical add-drop filter based on mismatched twin-core photosensitive-cladding fiber, *Opt. Lett.* Vol. 29, pp. 343-345.

Bhatia V. and Vengsarkar A.M., (1996) Optical fiber long-period grating sensors, *Opt. Lett.* Vol 21, pp. 692.

Davis, D. D., Gaylord, T. K., Glytsis, E. N. and Mettler, S. C. (1998) CO2-laser-induced long-period-fibre gratings: Spectral characteristics, cladding modes, and polarisation independence, *Electron. Lett.*, vol. 34, pp. 1416-1417.

Davis, D. D., Gaylord, T. K., Glytsis, E. N. and Mettler, S. C. (1999) Very-high-temperature stable CO2-laser-induced long-period fibre gratings, *Electron. Lett.*, vol. 35, pp. 740-742.

Dobb H., Kalli K., and Webb D. J., (2004) *Electron. Lett.* Vol. 40, pp. 657.

Eggleton B. J., Slusher R. E., Judkins J. B., Stark J. B., and Vengsarkar A. M., (1997) All-optical switching in long-period fiber gratings, *Opt. Lett.* Vol. 22, 883-885.

Erdogan T. (1997) Cladding-mode resonances in short and long-period fiber grating filters, *Journal of the Optical Society of America A*, Vol 14, pp 1760-1773.

Glogle D. 1971 Weakly guiding fiber, *Appl. Opt.* Vol. 10, pp 2252-8.

Grobnic D., Mihailov S. J., Smelser C. W., Becker M., and Rothhardt M. W., (2006) Femtosecond laser fabrication of Bragg gratings in borosilicate ion-exchange waveguides, *IEEE Photonics Technol. Lett.* Vol. 18, pp. 1403-1405

Han Y. -G., Kim S., and Lee S., (2004) Flexibly tunable multichannel filter and bandpass filter based on long-period fiber gratings, *Opt. Express* Vol. 12, pp. 1902-1907.

Hernández-García J.C., Estudillo-Ayala J.M., Rojas-Laguna R., Mata Chávez R.I., Martínez-Ríos A., Gutiérrez J.G., Trejo-Durán M., Vargas-Rodríguez E., Andrade-Lucio J.A. and Alvarado-Méndez E.(2010), Instrumentation and design of a high voltage source to produce optical fiber gratings by the electric arc technique, *Revista Mexicana de Física* Vol 56 pp. 255–2612.

Hill K.O., Malo B., Vineberg K.A., Bilodeau F., Jonson D.C., Skinner I.(1990) Efficient mode conversion in telecommunication fiber using externally written gratings, *Electron. Lett.* Vol. 26, 1270-1272.

Hill, K.O., B. Malo, F. Bilodeau, D.C. Johnson, and J. Albert. (1993). Bragg gratings fabricated in monomode photosensitive optical fiber by UV exposure through a phase mask, *Applied Physics Letters*. Vol. 62, pp. 1035-1037.

Hill, K.O., and G. Meltz. (1997). Fiber Bragg Grating Technology Fundamentals and Overview. *Journal of Lightwave Technology*. Vol. 15, No. 8, pp. 1263-1276.

Hill, K.O. (2000). Photosensitivity in Optical Fiber Waveguides: From Discovery to Commercialization. *IEEE Journal on Selected Topics in Quantum Electronics*. Vol. 6, No. 6, pp. 1186-1189.

Humbert G., Malki A., Fevrier S., Roy P., and Pagnoux D., (2003) *Electron. Lett.* Vol. 39, pp. 349

Hwang, I. K., Yun, S. H. and Kim, B. Y. (1999) Long-period fiber gratings based on periodic microbends, Optics Letters, Vol.24 pp 1263-1265.

Kakarantzas G., Dimmick T. E., Birks T. A., Le Roux R., and Russell P. S. J., (2001), *Opt. Lett.* Vol. 26, pp. 1137.

Kashyap R., (1999) fiber Bragg grating, Academic Press.

Kersey, A. D., Davis M A, Patrick Heather J, LeBlanc M, Koo K P, Askins C G, Putnam M A and Friebele E J (1997) Fiber grating sensors *J. Lightwave Technol. Vol.* 15 pp. 1442–63.

Mata-Chávez R.I. , Martinez-Rios A., Torres-Gomez, I., Alvarez-Chavez, J.A., Selvas-Aguilar, R., Estudillo-Ayala, J. M (2008) Wavelength band-rejection filters based on optical fiber fattening by fusion splicing, *Optics & Laser Technology*, Vol. 40 pp. 671-675.

Mata-Chávez R.I. Martínez- Rios. A. Torres-Gémez I., Selvas-Aguilar R., Estudillo-Ayala J.M. (2008) Mach-Zehnder All- Fiber interferometer Using Two in-Series Fattened fiber grating , *Optical Review*, Vol. 15 pp 1-6.

Meltz, G., W. W. Morey, and W. H. Glenn. (1989). Formation of Bragg Gratings in Optical Fibers by a Transverse Holographic Method. *Optics Letters*, Vol. 14, pp. 823-825.

Mihailov S. J., Grobnic D., Huimin D., Smelser C. W., and Jes B., (2006) Femtosecond IR laser fabrication of Bragg gratings in photonic crystal fibers and tapers, *IEEE Photonics Technol. Lett.* Vol. 18, pp.1837-1839.

Myoungwon K., Dongwook L., Bum I. H. and Haeyang C. (2002). Performance Characteristics of Long-Period Fiber-Gratings Made from Periodic Tapers Induced by Electric-arc Discharge, *Journal of the Korean Physical Society*, Vol. 40 pp. 369-373.

Lin C. Y., Chern G. W., and Wang L. A., (2001), Periodical corrugated structure for forming sampled fibre Bragg grating and long-period fiber grating with tunable coupling strength, *J. Lightwave Technol.*, Vol. 19 pp.1212-1220.

Liu Y., Chiang K.S., Rao Y.J. , Ran Z.L., and Zhu T. (2007) Light coupling between two parallel CO_2-laser written long-period fiber gratings, *Optics Express*, Vol. 15, pp. 17645-17651.

Othonos, A. (1997) Fiber Bragg Gratings. *Reviews of Science Instruments*, Vol. 68, No. 12, pp. 4309-4341.

Rego G., (2006) Chapter 3: Mechanisms of formation of arc-induced long-period fibre gratings" in *Arc-induced long-period fibre gratings: fabrication and their applications in optical communications and sensing*, University of Porto.

Rego G, Marques P.V.S., Santos J.L., Salgado H.M. (2005) Arc-induced long period gratings. *Fiber Integrated Opt* , Vol. 24 pp. 245–59

Rego G. M., Santos J. L., and Salgado H. M., (2006) Opt. Commun. Vol. 262, pp.152

Rego G. (2010) Fiber optic devices produced by arc discharges, *J.Opt.* Vol. 12 pp. 113002.

Riant I. (2002) UV-photoinduce fibre grating for gain equalisation, *Optical Fiber Technology* , Vol. 8, pp. 171-194.

StephenW.J. and Ralph P. T. (2003) Optical fibre long-period grating sensors: characteristics and application Meas. Sci. Technol. Vol 14 pp R49-R61.

Su L., Chiang K. S., and Lu C., (2005), Microbend-induced mode coupling in a graded-index multimode fiber *Appl. Opt.* Vol. 44, pp. 7394-7402.

Udd, E. (1991) Fiber Optic Sensors: An Introduction for Scientists and Engineers. *New York:* ed., *John Wiley and Sons.*

Vengsarkar A. M., Lemaire P. J., Judkins J. B., Bhatia V., Erdogan T., and Sipe J. F., (1996) Long-period fiber gratings as band rejection filters, *J. Lightwave Technol.*, vol. 14, pp. 58–65.

Vengsarkar A. M., Pedrazzini, J. R., Judkins, J. B., Lemaire, P. J. and Bergamo, N. S. (1996), *Optics Letters.*, Vol 21, pp 336.

Von Bibra M. L., Roberts A., and Canning J., (2001) Fabrication of long-period fiber gratings by use of focused ion-beam irradiation *Opt. Lett. Vol. 26, pp.765*

Wang Y.P., Xiao L.M., Wang D. N., and Jin W., (2007) In-fiber polarizer based on a long-period fiber grating written on photonic crystal fiber *Opt. Lett.*, vol. 32, pp. 1035-1037

Yablon A. D. (2005) Optical Fiber Fusion Splicing. Springer. Chap 8. 226

Fibre Sensing System Based on Long-Period Gratings for Monitoring Aqueous Environments

Catarina Silva[1], João M. P. Coelho[1], Paulo Caldas[2,3,4] and Pedro Jorge[3]

[1]*Departamento de Física, Faculdade de Ciências da Universidade de Lisboa*
[2]*Departamento de Física, Faculdade de Ciências da Universidade do Porto*
[3]*Unidade de Optoelectrónica e Sistemas Electrónicos, INESC Porto*
[4]*Escola Superior de Tecnologia e Gestão de Viana do Castelo*
Portugal

1. Introduction

The estuaries and coastal environments are areas of special biodiversity, being extremely sensitive to changes in its boundary conditions, particularly the properties of water, being salinity one of the most important parameters.

Thus, environmental monitoring is a crucial factor in the ecosystems health. Actually, methods for measuring salinity are mainly based on the mobility of ions in water (Zhao et al., 2003) changing the refractive index. However, these methods are inherently electrical and can be affected by any kind of electromagnetic interference. An alternative is the use of sensors based on optical technology, in particular, long-period fiber gratings (LPG) which are intrinsically sensitive to the external refractive index. These gratings devices have the advantage of being immune to electromagnetic interference and having small size, non-intrusive nature, flexible geometry, high speed of response and excellent salt water corrosion resistance.

Long-period gratings can be considered a special class of fiber Bragg gratings in which the period of the index modulation is such that it satisfies a phase matching condition between the fundamental core mode and a forward propagating cladding mode of an optical fiber. The periodic modulation of the refractive index in the fiber core typically has a period in the region from 100 μm to 1000 μm and a length of a few cm (James & Tatam, 2003) which can be induced in the fiber using different process. Irradiation by ultraviolet (UV) source, electric-arc discharges, irradiation with CO_2 source and application of mechanical pressure with adequate periodicity are the most common process.

The first LPG successfully inscribed in an optical fiber was described in 1996 by Vengsarkar (Vengsarkar et al., 1996a, 1996b) and was used as a band-rejection filter. In the same year Bhatia (Bhatia & Vengsarkar, 1996) presented the first LPG device acting as a sensor. Since then, many applications for long-period gratings have been achieved in the last years and have been concentrated on the development of long-period grating-based fiber devices for use in optical communications and fiber optic sensor systems. In optical communication systems, LPGs are applied as gain equalizing filters, wavelength selective devices, bandpass and band rejection filters and wavelength tunable add/drop multiplexers (Vengsarkar et al., 1996a, 1996b; Dong et al., 2009). In the field of sensing systems, they can be applied as structural bend sensors, temperature sensors, axial strain sensors, refractive index sensors and biochemical optical sensors (Patrick et al., 1998a; James & Tatam, 2002; Bhatia, 1999; Falciai et al., 2001).

The high sensitivity of LPGs to the refractive index of the surrounding environment allows them to be used as salinity sensors in aqueous media. However, in their production the removal of the buffer layer makes them a fragile structure when applied in real conditions, usually contaminated with algae and other organic materials. Taking this into consideration, this chapter addresses the potential application of long-period gratings technology in environmental monitoring. It will analyze the response of different gratings to refractive index changes, in order to evaluate and increase their sensitivity, as well as its thermal response. In this chapter, we present the issues related with LPG implementation in a real field application and review the use of LPG based sensing heads for environmental monitoring in field conditions. As a particular example, a prototype of a salinity sensor developed by the authors will be described for in situ research and capable of sustaining harsh operational conditions.

The chapter starts by presenting the main theoretical concepts related with LPGs, focusing on their fundamental aspects and fabrication methods and also its application as sensor elements (section 1). In section 2, the main configurations of refractive index sensors based on LPGs are reported. The problematic of encapsulating the sensing element and make it a robust sensor device will be addressed, in section 3. Mechanical requirements are described and the filtration issues analysed. The easiness of device cleaning is another crucial aspect that will be taken into consideration. Finally, in section 4, a prototype of a refractive index sensor for in situ research will be proposed and characterised and the potential of the sensing system as a remote refractometric device is analysed.

1.1 Fundamentals of long period gratings

As mentioned before, in 1996 (Vengsarkar et al., 1996a) it was reported the principle of operation of an LPG as a device based on the coupling between the fundamental mode propagating in the fiber core with the forward propagating cladding modes. This principle is illustrated in Fig. 1. The cladding modes are quickly attenuated and this results in series of loss bands in the transmission spectrum of the grating. Each of these attenuation bands corresponds to coupling to a discrete cladding mode.

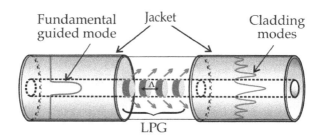

Fig. 1. Coupling of fundamental guided mode to a cladding mode in a long period grating.

The resonant wavelength of light coupling into a particular cladding mode is given by the phase matching condition (Erdogan, 1997a, 1997b):

$$\lambda_{res}^{m} = \left(n_{eff,co} - n_{eff,cl}^{m} \right) \Lambda \tag{1}$$

where Λ is the grating period, $n_{eff,co}$ and $n_{eff,cl}^{m}$ are the effective refractive indexes of the core and mth-cladding modes, respectively.

The light coupled into cladding-guided modes is, most typically, absorbed in the fiber buffer or radiates away from the fiber. Thus, this type of fiber grating acts as a wavelength selective transmission filter. The wavelength transmission response of a typical LPG is shown in Fig. 2, exhibiting the characteristic attenuation bands.

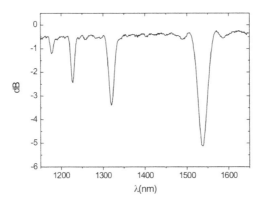

Fig. 2. Transmission response of a typical LPG.

The light transmission through the core follows a sinusoidal function of the core refractive index modulation for the wavelengths in the resonance (Zanlorensi et al., 2009)

$$T = \cos(DL/2) \tag{2}$$

where L is the grating length and D is a coupling coefficient proportional to the core index modulation. The bandwidth of the resonance dips depends on both the coupling coefficient and the difference between the core and cladding indexes:

$$\Delta \lambda_{FWHM} = \frac{\lambda_{m}^{2}}{\left(n_{eff,co} - n_{eff,cl}^{m} \right)} \sqrt{\frac{4D}{\pi L}} \tag{3}$$

External changes in parameters like refractive index, temperature or strain can affect the terms in Eqs. (1) and (3), consequently shifting the attenuation dips and altering their bandwidths (Zanlorensi et al., 2009).

1.2 Fabrication techniques

LPGs are created by inducing a periodic refractive index modulation in the core of the optical fiber with period lengths of typically several hundred micrometers. This can be made by permanent modification of the refractive index of the optical fiber core or by the physical deformation of the fibre. Since the first demonstration of LPGs by writing the grating with UV laser light through an amplitude mask in 1996 (Vengsarkar et al., 1996a), several methods have been developed to fabricate theses structures. The most widely used method is the exposition of the photosensitive optical fibres to UV light through an amplitude mask, phase mask or by interferometry (Bhatia & Vengsarkar, 1996; 1998a, James & Tatam, 2002; Erdogan, 1997a). In this case, the mechanism responsible for the refractive index change is related with the formation of Germania-associated defects and the gratings can be written using light with wavelengths between 193 and 355 nm (Bhatia & Vengsarkar, 1996; Zanlorensi et al., 2009). The amplitude mask technique consists of simply imprinting the mask pattern onto the core of a photosensitive optical fiber placed in contact with or in close proximity immediately behind the mask. The fiber exposure to the fringe pattern induces a refractive index modulation in the core with the same periodicity as that of the mask (Nikogosyan, 2006). The mask might be fabricated in chrome-plated silica or metal foil. The disadvantages of this method are that the grating period and length are restricted by the dimensions of the amplitude mask which can also be easily damaged if exposed to UV light whose intensity exceeds its damage threshold. A further useful technique was presented by Hill et al. in 1990 (Hill et al., 1990) in the form of point-by-point periodic grating manufacture. In this technique, gratings are written period after period, and a single pulse of UV light from an excimer laser passes through a mask containing a slit making the refractive index of the core in the irradiated fiber section locally increase. However, all process that depends on UV illumination requires first that the fiber must be receptive to this electromagnetic radiation. In other words, the fiber must be made photosensitive before writing the grating. This can be achieved by doping the fiber silica core with impurity atoms (such as germanium, boron) or by hydrogen loading (Kashyap et al., 2010, Lemaire et al., 1993).

Kondo et al. (Kondo et al., 1999) found that focused irradiation of femtosecond laser pulses causes a permanent refractive index increase in various glasses, and they demonstrated a novel technique for the fabrication of long-period gratings by focused irradiation of femtosecond laser pulses. They induced a refractive index periodic structure in the core of a single-mode fiber by using light that is not absorbed by the core and cladding glasses and a polymer coating. In contrast with the UV-laser induced, LPGs fabricated using infrared (IR) femtosecond lasers have high resistance to thermal decay despite having no special stabilization.

Another possibility to create these structures is by applying the ion implantation technique. This technique induces a refractive index increase in silica-based glass, which is mainly due to the densification of the silica glass induced by atomic collisions (Fujimaki et al., 2000). Disadvantages of LPG manufacture by using this technique includes the necessity of specialized equipment (such as a tandem accelerator), as well as the fact that the cladding material average refractive index is also raised in the process, which leads to large background losses in the transmission spectrum. The main advantage of ion implantation is that it is possible to inscribe LPGs in almost any type of silica-based fiber without prior photo-sensitization.

Gratings inscription was also achieved through the use of thermal effects by exposure to near-IR carbon-dioxide (CO_2) laser light (Davis et al., 1998; Kim et al., 2002) or electric arc discharges (Rego et al., 2001; Falate et al., 2006). Both techniques allow the inscription of the gratings without the need hydrogen-loaded or Ge-doped core fibres and are more flexible than the UV technique since it is not required the fabrication of a specific amplitude mask for each desired resonance wavelength (Humbert & Malki, 2003). The first results of the CO_2 laser application in the fabrication of LPG on fiber optics were published in 1998 (Davis et al., 1998; Akiyama et al., 1998; Liu et al., 2011). In this case, the residual stress relaxation and glass densification are usually considered to be the main mechanisms responsible by the formation of LPGs and have demonstrated very high thermal stability. In this technique, the 10.6 µm wavelength laser radiation is focused to a spot by means of spherical lenses and LPGs are written by periodical local heating. The grating period is determined through the movement of a computer controlled translation stage. During the writing process, a broadband source and an optical spectrum analyzer are used to measure the transmission spectrum.

The other process used to write LPGs by periodical local heating of the fiber is based on electric arc-discharges. In this technique the refractive index modulation results from deformation of fiber due to intense heating caused by an electrical discharge transversal to the fiber. This writing method is very flexible, easy to implement and does not require the use of special fibres or high cost equipment, such as UV or IR lasers. These features make this technique particularly suitable for the low cost fabrication of LPGs. An example of a typical electric arc-discharge setup for LPGs manufacture is schematized in Fig. 3. The fabrication process consists of positioning an uncoated fiber between the electrodes of a commercial splicing machine. A small weight is suspended in one of the fibre's extremities to keep a constant longitudinal tension. The other extremity of the fiber is mounted on a computer controlled translation stage. An electrical arc discharge is applied with the fusion splicer, using an electric current of 8.5–10.0 mA for 0.5–2.0 s. A translation stage moves the fibre, after each discharge, by a distance that represents the grating period (typically 300–700 µm). The refractive index change is created by the fast heating and cooling discharge cycles. Due to the asymmetric nature of electric arcs, the result of each consecutive discharge is

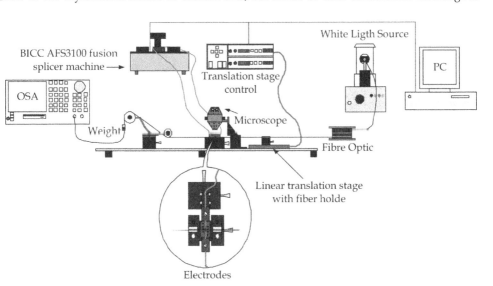

Fig. 3. Schematic diagram of LPFG fabrication system based on electric arc-discharge.

never exactly the same. Therefore, in order to obtain an LPG with a given characteristics, the evolution of the transmitted spectrum is monitored by an optical spectrum analyser and a White Light Source (WLS). When the measured spectrum shows suitable characteristics to the intended application, the process is terminated. Usually 15–100 points are necessary to form the grating.

Since LPGs have periods of the order of hundreds of micrometers they can also be induced mechanically through periodic microbends in optical fiber (Rego et al., 2003; Yokouchi et al., 2005; Cardenas-Sevilla et al., 2009) Several configurations have been developed to produce mechanical gratings but they share the same objective, that is, to induce periodic stress and/or microbending in a single mode fibre (Rego et al., 2003). This technique, as the arc discharge technique, can potentially be applied to any kind of fibre and is also simple, flexible and low-cost. Another important advantage over other techniques is the fact that these gratings can be produced without removing the fibre coating and therefore the fibre keeps its mechanical integrity.

1.3 Sensing characteristics

As mentioned before, LPGs are influenced by alterations that can occur in the environment at the grating location, such as changes in the refractive index or temperature, resulting in measurable changes in the LPG transmission spectrum. This is the basis of the sensing LPGs (Bhatia, 1999). In this section, it is presented an analysis of the implementation of LPGs as an element sensor of physical parameters such as temperature, mechanical deformations and refractive index of the environment.

1.3.1 Temperature sensitivity

The temperature sensitivity of the LPGs may be understood by differentiating the phase matching condition equation (Humbert & Malki, 2002) Since the contribution of the waveguide property $(d\Lambda/dT)$ to the resonance wavelength shift with temperature variation is negligible compared to that of the material property (dn/dT) in LPGs the derivative of equation (1) can be set as

$$\frac{d\lambda_{res}^m}{dT} = \frac{d\lambda}{dT}\left(n_{eff,co} - n_{eff,cl}^m\right) + \Lambda\left(\frac{n_{eff,co}}{dT} - \frac{n_{eff,cl}^m}{dT}\right) \approx \Lambda\left(\frac{n_{eff,co}}{dT} - \frac{n_{eff,cl}^m}{dT}\right) \quad (4)$$

Fig. 4 shows the wavelength response and the spectral shift of the LPG (induced in Corning SMF-28 fiber) when subjected to temperature variation from 0 °C (related to a 20 °C ambient temperature). As expected, the resonance peaks shift to longer wavelengths with increasing temperature. The heating causes an increase in the difference between effective refractive indexes of the modes due to the thermo-optic coefficient of the core being larger than that of the cladding, causing a shift in wavelength to higher values. As a coarse approximation, a linear fit to the wavelength shift data yields a slope of ~0.085nm/°C.

1.3.2 Refractive index measurement

An unusual feature of LPGs is their high sensitivity to the refractive index changes in the grating location. This alteration will also change the matching condition expressed by

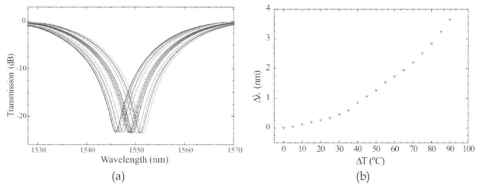

(a) (b)

Fig. 4. (a) Transmission spectrum and (b) shift in the wavelengths of the LPG with respect to temperature variation.

equation (1) and will lead to wavelength shifts of the resonance dips in the LPG transmission spectrum. Such displacement occurs due to the dependence of the effective refractive index of the cladding modes on the refractive index of the surrounding medium. The sensibility to refractive index changes increases with the reduction of the difference between the two effective refractive indexes.

Fig.5 shows an example of the evolution of the LPG transmission spectrum when the refractive index of the external medium changes. For calibration purposes, the LPGs were immersed in liquids with different refractive indexes in the range [1.33, 1.43] at room temperature in both situations. It can be seen that as the refractive index of the external medium increases the attenuation peak shifts towards shorter wavelengths. This is an expected displacement because an increase in the effective refractive index of the cladding according to the resonance condition leads to lower resonant wavelengths (Rego et al., 2006).

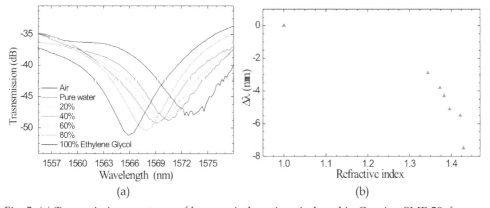

(a) (b)

Fig. 5. (a) Transmission spectrum of long period gratings induced in Corning SMF-28, for different refractive index and (b) corresponding variation in the shift of the resonant wavelength.

The sensitivity of LPGs to refractive index and temperature is influenced by the composition of the optical fiber. Rego (Rego et al., 2006) investigated the sensitivity of arc-induced LPGs in pure-silica-core fibres to refractive index changes of the external medium. These results show a two to three times increase of the sensitivity when compared to gratings written in standard fibres. This results from doping the cladding with fluorine which lowers its refractive index and, therefore, reduces the refractive index between the cladding and its surroundings, increasing the grating sensitivity.

As the effective refractive index of the cladding modes is strongly dependent on the cladding diameter, the easiest way to increase the sensitivity of the LPGs is by etching the fiber cladding in the region of the grating to a diameter such that the evanescent field of the mode displays a higher overlap with the immediate surrounding environment. This concept is clearly illustrated in the phase-matching condition (equation 1), where for a fixed period the change in effective refractive index for a higher order m of the cladding mode results in a new resonant wavelength - the effective refractive index of the core remains unaffected. With this configuration, the value of the effective refractive index of the waveguide is directly affected by the refractive index of the medium, and this effect is stronger for higher order modes.

2. Refractometers based on long period gratings

In recent years the development of devices for measuring the refractive index has been the subject of great interest for applications in industrial processes and quality control, biomedical analysis, and environmental monitoring. In this context, the LPGs offer advantages, being extremely sensitive to changes in refractive index of the medium where they are immersed. However, although the use of such devices is advantageous, its use as a sensing element is limited by the relatively large rejection bandwidth (~10 nm) which does not provide a good resolution. In this context, several schemes for refractive index sensing based on LPGs have already been proposed (James et al., 2002; Han et al., 2001).

The simple configuration, consisting of a light source, an LPG immersed in a fluid and an optical spectrum analyzer has been studied by several authors (Falciai et al., 2001; Zhu et al. 2003). Lee et al. (Lee et al., 1997) demonstrated a new analysis method to determine the spectral shift of an LPG as a function of refractive index of the medium. The change of this parameter with the wavelength of the LPG attenuation band was presented by Patrick (Patrick et al., 1998a). As previously mentioned, the sensitivity of the sensor to external refractive index change can be enhanced by decreasing its cladding diameter by chemical etching of the optical fiber (James & Tatam, 2003). The reduction in cladding diameter changes the effective index of the cladding modes, resulting in an increase in the central wavelengths of the attenuation bands.

Gwandu et al. proposed a compact scheme for simultaneous measurements of temperature and refractive index in a medium, using two LPGs recorded in a double cladding hydrogenated fiber (Gwandu et al., 1999). The effect of the thickness of layers deposited on the grating surface regarding its response to refractive index changes has also been investigated (e.g. James et al., 2002). A thin film of organic material was deposited on an LPG through the Langmuir-Blodgett technique. The central wavelength and the minimum transmission band attenuation of the LPG depend on the thickness and refractive index layer of the deposited material, even when the refractive index is higher than the cladding. It was observed that for films with refractive index higher than that of the cladding, the wavelength and amplitude attenuation of the band exhibits a very high sensitivity when the film thickness is of the order of a few hundred nanometers. If the material deposited has a refractive index below that of the cladding, the sensitivity to film thickness is considerably reduced.

LPG pairs have shown high resolution to refractive index measurement compared to the use of a single LPG (Han et al., 2001; Zhang et al., 2005). The advantage of using a LPGP sensor relies upon its principle of operation, where the coupled core and cladding modes from the first LPG combine again at the second matched LPG to form interference fringes. The core and cladding paths constitute the arms of a Mach–Zehnder interferometer (MZI) in the transmission mode (Mishra et al., 2005; Murphy et al., 2007) or a Michelson interferometer in the reflection mode (Swart, 2004). A schematic diagram of these two configurations is presented in Fig. 6.

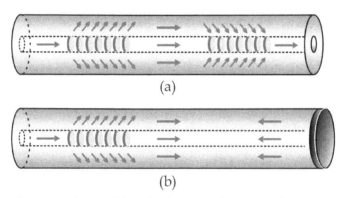

(a)

(b)

Fig. 6. Schematic diagrams of an LPG-based: (a) Mach-Zehnder interferometer; (b) Michelson interferometer.

The Mach-Zehnder interferometer operate in transmission mode and is obtained by placing in series two identical LPGs inscribed adjacent to one another such that an interferometer cavity of a certain length is formed between the two gratings. When light from a broadband source passes through the first LPG, the part that propagates in the fundamental mode is coupled to the forward propagating cladding modes. At the second LPG, the cladding mode is coupled back into the core. Due the difference between the effective refractive indexes of the core and cladding modes, the light coupled into the core by the second LPG is phase shifted with respect to the light that propagated through the core, giving rise to the interference fringes (Lee & Nishii, 1998). The core and cladding paths constitute the arms of the Mach–Zehnder interferometer. The frequency of the fringes depends on the separation between the two gratings, while the phase is influenced by measured-induced changes in the effective index of the cladding modes (Murphy et al., 2007). The fine interference fringes in the transmission spectrum of the pair of LPGs can enhance the resolution and the sensitivity of measurement. The narrower bandwidth of the fringe facilitates greater resolution in the measurement of the wavelength than is possible with conventional LPGs (Han et al., 2001).

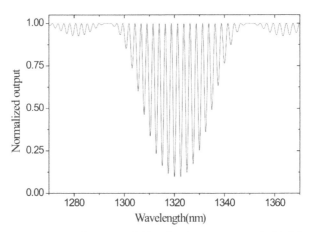

Fig. 7. Transmission spectrum of an optical fiber having two cascaded LPGs as a Mach-Zehnder interferometer.

The in fibre-Michelson interferometer was proposed and studied by Swart (Swart, 2004). This configuration is based on a single LPG and on a fibre end mirror, as show in Fig. 6.b). In this case, the light coupled to cladding modes is reflected by the mirror and finally recoupled back to fibre core. This is a very simple structure and offers several advantages over the Mach-Zehnder configuration such as increased sensitivity, operation in reflection mode and lower manufacture complexity (it avoids the need to fabricate identical LPGs). This reflection mechanism and compact size make this kind of sensing-structure attractive as biomedical probes.

The misaligned splicing point (MSP) formed by a commercial fusion splicer may replace the first LPG in an LPG pair to create an MZI (Dong et al., 2005). In this, the coupling of light from the core to the cladding modes is made trough the MSP, as illustrated in Fig. 8.

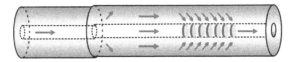

Fig. 8. Misalignment splicing point is formed to interact with the single-mode fibre-based LPG

Ding et al (Ding et al., 2005) reported the fabrication and characterization of a sensor with high sensitivity to refractive index changes based on a couple of LPGs in which it was reduced the fibre section between them (FTS-LPGP-fibre-taper seeded LPG pair). Its principle of operation is schematised in Fig. 9. This configuration allowed an increase by a factor of five in the sensitivity to refractive index change on the external environment when compared to the conventional configuration (Frazão et al., 2006).

Allsop et al. (Allsop et al., 2006), demonstrated that a Tapered-LPG configuration has a high spectral sensitivity to refractive index change in liquids. This configuration is based on the writing of a LPG centred on a taper.

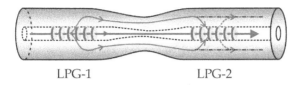

LPG-1   LPG-2

Fig. 9. Schematic diagram of a FTS-LPGP (fiber-taper seeded LPG pair).

Recent advances have been focused in the research of LPGs induced in photonic crystal fibres (PCFs) which consist of a pure-silica core, and a microstructured air-silica cladding (Yu et al., 2006). Because the microstructured cladding consists of air holes, a PCF's cladding index shows strong wavelength dependence. This makes such fibres exhibiting some properties that make them advantageous in face to conventional fibres (e.g. by reducing or even eliminating the cross-section problems).

3. Sensing head

Several authors have experimentally demonstrated different configurations for the application of LPGs as sensor elements. However, when the LPGs are to be deployed in harsh environments, their fragile nature makes the existing technology not satisfactory. One of the harshest environments where fiber-based sensors have a huge potential are natural aqueous media. The existence of biological activity is added to other debris that can not only alter the sensing capability but severely damage the entire sensing head.

This section addresses the several issues that must be considered in the project of a system to host a LPG in order to reduce the potential cross-sensitivity problems and obtain a high performance when applied in real conditions in natural aqueous environments. The issues related to the mechanical requirements in order to achieve a high performance sensor device under the considered conditions will be analysed. Strain constrains, supporting conditions and other operational prescriptions will be addressed. Also, the characteristics of a system developed by the authors will be described in detail.

3.1 Mechanical requirements

The mechanical part of the sensor head should be projected in order not only to bear the sensor element, but also give support to filtering elements (latter addressed) and allow efficient cleaning procedures, part of the unavoidable regular maintenance.

3.1.1 Structure

First considerations should regard the material used to produce the supporting structure. Being in direct contact with the aqueous medium this choice is of prime importance for any project. Resistance to corrosion and structural rigidity are the main characteristics to be taken in consideration, while price and machinability are also usually considered. Two main groups can be considered: metals (including alloys) and polymeric materials.

When metals and alloys interact with an environment, their behaviour is highly dependent on both the thermodynamics and kinetics of the interaction. The chemistry, solubility, structure, thickness, cohesion, adhesion, continuity, and mechanical properties of the reaction compounds contribute to the formation of either passivation or corrosion, two concurrent processes. In fact, passivation can be defined as the spontaneous formation of a hard non-reactive surface film inhibiting further corrosion. This layer is usually an oxide or nitride a few nanometres thick.

Contrary to metals, polymers are not affected by corrosion processes although susceptible to light (mainly, ultra violet) and potentially less structurally stable. Nevertheless, the research in engineering polymers is highly advanced and application-dedicated materials can be developed. So, depending on overall functional requirements, a wide range of materials can be chosen. The structure and components of the sensing head may use carbon steels, low alloy steels, aluminium alloys, copper and copper alloys, stainless steels, superaustenitic stainless steels, nickel-base alloys, cobalt-base alloys, ultra-high strength steels and stainless steels, titanium and titanium alloys, polymers, or composites. Newer materials such as intermetallic compounds, nano-structured alloys, and amorphous alloys are promising materials, but their collective database of respective properties is currently extremely limited (Shifler, 2005).

Choosing the correct material in the subject of this paper depends on very specific data regarding the application. If, as mentioned, the environment is a key point to consider, the kind of measurements to perform, duration of testing, production numbers (mass or small production) and available budget easily superpose. At the end, nowadays there are a sufficient number of materials that can perform well in harsh environments.

3.1.2 Tensing element design

Since the LPGs are intrinsically sensitive to bending curvature, which consequently influences its sensitivity to the various parameters, for field deployment in practical applications, it becomes necessary to design a support system that can avoid these cross sensitivities. A particular solution tested by the authors was a spring-based system designed in order to apply a constant force to the grating regardless of the external handling conditions. In general, the design of a compression spring must consider the following factors:

1. space for housing the spring;
2. rigidity requirements;
3. guidelines for the values of displacement and maximum force regarding the values imposed by the project.

Fig.10 shows a helical spring subjected to a compressive deformation, y, where d is the diameter of wire and p the helix pitch (Shigley, 1986).

The elongation (or contraction) of a spring is determined by the torsion deformation of all the active coils of the spring. Then, the deformation of the spring coils with N_a active coils under the action of a force P is given by

$$y = \frac{8PD_m^3N_a}{d^4G} \qquad (5)$$

being G the material's elasticity modulus and D_m the mean diameter or the coils.

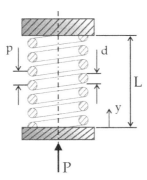

Fig. 10. Helical string subjected to compression.

The helical spring constant or stiffness, κ, which reflects the amount of deformation of a spring, y, when subjected to a load P, can be determined by,

$$k = \frac{P}{y} = \frac{d^4 G}{8 D_m^3 N_a}$$ (6)

Representing the length of the unloaded-spring by L_0, the helix step, p, is

$$p = \frac{L_0}{N_a + 1}$$ (7)

and considering the spring ends as having close spirals, the total number of coils, N_t, is given by

$$N_t = N_a + 2$$ (8)

3.2 Filtering system

Essential in the sensing head performance is the encapsulation process in view of the specifications of a filtration system. In real conditions, the system hosting the LPGs requires an adequate filtering in order to prevent the deposition of organic materials and the accumulation of particles around the sensor element, hence interfering in the measured value. The filtration process to protect the sensing head is essentially based on the passage of fluid through a porous or permeable material. The encapsulation proposed for the sensor involves two levels of filtration: one capable of retaining larger particles and another for the mud usually existing in natural aqueous environments.

For filtrating larger particles it is common the use of lower pore sizes in order to prevent the deposition of waste or algae. The filtration of mud is achieved through the process of membrane separation. This consists in running a liquid through a membrane, which acts as a selective barrier and when subjected to pressure, forced or natural, leads to the passage of fluid and smaller compounds through their pores.

The membranes can be manufactured with organic or inorganic material. Inorganic membranes exhibit higher mechanical strength, thermal and chemical robustness and have a longer active life. The inorganic membranes can be ceramic (zirconium oxide, aluminium or titanium), metal and graphite. The efficiency of the membrane depends on several factors, such as the properties of membrane and fluid and operating conditions (pressure, temperature, turbulence) (Queiroz et al., 2001).

Their morphology is mainly related with the density and porosity (number and dimension of the pores) of the layers constituting the membrane. The porosity, thickness, pore size and permeability are important factors in the choice of the membrane. These characteristics depend on the material of which it is made off and on the manufacturing technique. The higher the porosity the lower the resistance to the fluid flow through the membrane is. In general, the classification of the membranes is based on the structure and mechanism of separation. Regarding the structure of a membrane, it may be porous or nonporous, and symmetrical or asymmetrical, as is illustrated in Fig. 11. The anisotropic membranes are characterized by presenting a very thin, more packed, layer supported by a porous layer. In the case of porous membranes, the pore size and its distribution will determine which particles or molecules will be retained by the membrane (Queiroz et al., 2001).

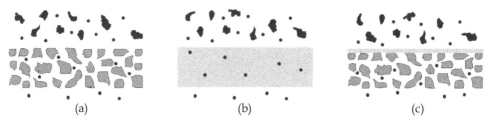

(a)   (b)   (c)

Fig. 11. Schematic representation of (a) symmetric porous and (b) non-porous membranes and (c) asymmetric porous membranes.

Among the existing membrane separation processes, we only address the micro-and ultra-filtration processes. These processes differ among themselves mainly in the pore size and pressure required for operation.

Microfiltration can be defined as a process for solid-liquid separation using membranes with pore diameters ranging from 0.1 μm to 2.0 μm, and operating at pressures between 0.2 bar and 2.0 bar depending on the product to be filtered. The filter is a micro-porous membrane and this technique allows you to retain suspended particles, bacteria and some ions (Queiroz et al., 2001).

The ultrafiltration process operates at pressures between 2.0 bar and 7.0 bar and uses membranes with micro-pores of 1 nm to 100 nm, which allows the passage of small molecules (water, salts) retaining all suspended solids including those causing turbidity and even micro-organisms. The flow through the membrane is controlled by the pressure exerted on the liquid. The main limitation of this process is the membrane blockage, which occurs when particles are deposited on the filter inside leading to a decrease in its filtration capacity. The choice of a membrane with pores smaller than the diameter of the particle filter and subjected to a pressure gradient, prevents this phenomenon. Operating conditions such as flow, velocity and pressure concentrated in the pores are also factors that influence the degree of clogging of the membranes. The application of a pressure gradient in the membrane in contact with the solution promotes a flow of fluid through the membrane.

The main advantages of membrane separation processes are:

1. high selectivity by using one process or the coupling of processes;
2. generally, these processes operate at room temperature, it is not necessary to change the fluid temperature to promote the phase separation;
3. low power consumption: promoting the separation that occurs without phase change;
4. easy to combine with other processes.

The main disadvantage is its high cost. However, costs associated with the application of this technology have been reduced considerably, since the membranes are produced on a larger scale, there is a greater number of firms in the market and, when properly applied, the membranes have a longer service life and flow permeated most stable and high.

3.3 Prototype

The guidelines explained in the previous text were applied in the design of a mechanical device with the functions of supporting and housing a sensor for measuring salinity and temperature, which is based on a LPG inscribed in standard silica optical fibre, with 125µm in diameter. The developed system is intended to operate at half depth in the Ria de Aveiro, inserted in the optical cable, protecting the optical fibre (Silva et al., 2010).

3.3.1 Mechanical device

Being a prototype and not expected to operate for longer periods of time the main driver for the choice of the material was cost an easy of manufacture. For this reason, common aluminium (1060 Alloy) was chosen.

A cylindrical shape was chosen in order to allow the insertion of an outer mud filtrating membrane and a protective mechanical grid. The sensor element is surrounded by enough space to promote the flow of water allowing updating salinity measurements. The optical cable is fixed at both ends by mechanically blocking the fiber at the head's support, leaving about 11 cm of exposed optical fiber. The device allows the adjustment of the tension in the fiber through a spring. Fig. 12 presents a picture of the prototype.

3.3.2 String dimensioning

The coil spring used in the device was scaled by imposing a force (compression) of 20 g over a diameter of 40 mm. The spring, with a free length of 130 mm, was made of stainless steel which has an elasticity modulus of 80 GPa and a wire diameter of 1mm. The number of active coils necessary can be calculated from the equation (5). So, the number of active coils is 14. Assuming squared and ground ends, the total number of the coils is 16 (equation 8).

3.3.3 Filtering system

The selection of the membranes used in this work was done taking into account the environment where they are to be inserted and through well-defined parameters such as its material, thickness and pore size. Three types of membranes were analysed with application in aqueous environments, two with the Nylon and the other with fibreglass mesh. Its specifications are shown in Table 1.

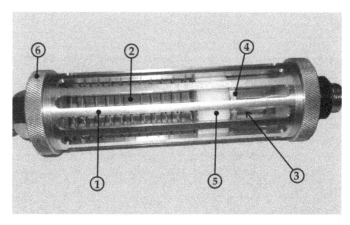

1. Cylinder (membrane support)   4. Plunger
2. Spring (control fiber strain)   5. Ring small bolt
3. Semi-blocker of optical fibre   6. Ring bolt-fixing

Fig. 12. Picture of the mechanical device developed to support the element sensor.

	Mesh thickness [μm]	Pore dimension [μm]	Maximum temperature [K]	
			(dry environment)	(wet environment)
Fiberglass	0.52	2.7	-	-
Nylon 53	60	53	388	
Nylon 100	78	100	373	

Table 1. Specifications of the analyzed membranes (Cole-Parmer instrument Co).

The analysis to establish the most suitable of these filters was made *in situ*. Supports were projected based on a net applied surrounding two metallic plates, and over which the membrane to test is fixed (Fig. 13). All metallic parts were designed in aluminium. After placing the membranes in the medium, they were immersed in the Ria de Aveiro, Portugal, where they remained for one month. After this period of time, the membranes were analysed regarding their state of blockage using an optical microscope (Meiji). Inside the vessel, another membrane was placed and a similar analysis was made in order to have an idea of the amount of impurities that have passed the outer filter.

In Fig. 14, it can be observed the state presented by the membranes after the specified period under real conditions. It was found that the fiber glass membrane did not supported the conditions of the environment concerned, and may well be conclude that it is not suitable for the application in question. It can also be seen that for the same period, the nylon membrane with pore size of 53 μm shows greater accumulation of dirt compared to the nylon membrane pore with a size of 100 μm, as expected. Fig. 15 shows the microscope images for both nylon membranes. The nylon membrane with larger pore sizes was fully saturated, while the other still had some open pores. However, because of its condition, it appears that the replacement, in real operation, has to be done in less than a month. The nylon membrane with pore size of 53 μm placed inside the support was also relatively clean.

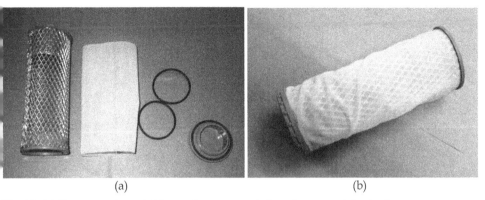

(a)   (b)

Fig. 13. (a) Components and (b) membrane mounted on its support for testing.

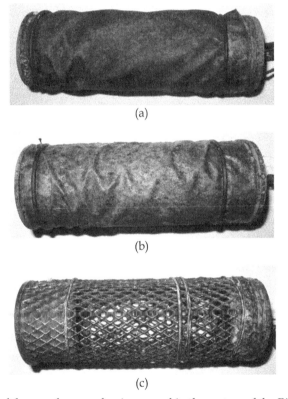

(a)

(b)

(c)

Fig. 14. Condition of the membranes after immersed in the waters of the Ria de Aveiro, Portugal, for one month: (a) Nylon (100 μm), (b) Nylon (53 μm) and (c) Fiberglass.

According to the results presented we can conclude that for the application in question nylon fiber with 53 μm pore size is the one that best meets the requirements. However, its replacement should occur in relatively short periods (at least every two weeks).

(a)   (b)

Fig. 15. Images obtained from optical microscopy of the Nylon membranes, (a) 100 μm and
(b) 53 μm, after immersed in the waters of the Ria de Aveiro, Portugal, for one month.

These tests also allowed analysing the performance of an aluminium structure subjected to
the aqueous harsh conditions of Ria de Aveiro. In spite of its susceptibility to corrosion, the
structure supported relatively well the one month "ordeal". However, the existing screw
threads were clearly damaged, showing a tendency to weld the parts in contact.

4. Sensing head performance

The performance of the previously defined system is analysed in this section. In order to
evaluate the influence of the structure on the measurement of environmental parameters, a
comparative study is made when the LPG is fixed with and without the structure. The
sensitivity of the sensing head to the refractive index change of the external medium was
characterized. The thermal response of the sensing structure was also studied. The LPG
used in this work were written by electric arc technique (Rego et al., 2005) in a standard
telecommunication fibre (Corning SMF-28), with a period of ~ 540 μm and a length of ≈ 2
cm. To further enhance its sensitivity for refractive index changes, the fibre diameter was
reduced. This was made by chemical attack with aqueous solution of fluoridic acid (HF
40%). After the etching process, the fibre was thoroughly washed with deionised water,
dried, and observed under the microscope. The diameter of the fibre after suffering the
chemical etch was 118.8 μm ± 1 μm, considering an initial diameter of 125 μm.

4.1 Refractive index measurement

To check the suitability of the system developed for salinity refractive index measurements,
the sensitivity of the LPG based sensing head was characterized towards changes in the
refractive index of the external medium. For calibration purposes, in both situations the
LPGs were immersed in liquids with different refractive indexes at room temperature. The
liquid was prepared by using a mixture of water and different ethylene glycol
concentrations were used, resulting in samples with refractive indexes in the range [1.33,
1.43]. The refractive index of the different samples was calibrated using an Abbe
refractometer. A broadband light source (Photonetics – Fiber White), with emission centred
on 1550 nm illuminated the sensing element. From the transmitted broadband spectrum, the

refractive index can be measured from the spectral shift of the LPG attenuation band. This shift was monitored using an optical-spectrum analyzer (ANDO AQ 6330). Fig. 16 shows the grating responses to the surrounding solutions with increasing concentration of ethylene glycol, with the device either free (without curvature) or integrated in the structure.

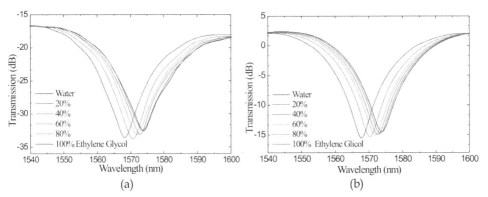

(a)   (b)

Fig. 16. Evolution of the transmission spectra of the LPG, (a) free and (b) mounted in the sensing head, as the refractive index is changed by increasing the concentration of ethylene glycol.

It can be seen that the resonance wavelength moves towards the shorter wavelength region as the concentration of ethylene glycol is increased. This can be understood through an increase of the effective refractive index of the cladding, which, according to the resonance condition, leads to lower resonant wavelengths (Lee et al., 1997). The graph presented in Fig. 17 derives directly from the analysis of the central peak of each attenuation band presented in Fig. 15. The shift is determined with reference to the spectral position considering air as the surrounding medium.

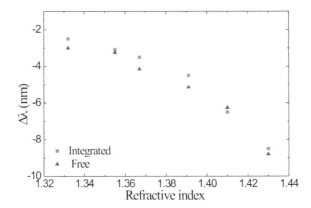

Fig. 17. Resonance wavelength shift of the LPG measured for different refractive indexes of the external medium in the situations where the device is free and integrated in the structure.

It can be seen that for both situations there is a similar response of the grating to refractive index changes. The slope of the obtained calibration curves gives an idea about the sensitivity of the refractometer. For a refractive index around 1.40, the sensitivity of the LPG is ≈ -67nm/RIU and -57nm/RIU for the cases where the device is mounted and free, respectively (RIU: refractive index unit). This difference is an indication that the pre-strain of the LPG in the structure induces a small enhancement of its refractive index sensitivity. As expected, it was also observed that the sensitivity increases substantially when the refractive index of the external medium is close to the cladding refractive index.

4.2 Salinity measurement

In order to analyse the influence of the salt concentration, the spectral responses of the LPG mounted in the structure was measured when it was immersed in water with different salt concentrations, resulting into salinity levels between 1% and 19%. The spectral response evolution of the LPG when mounted in the structure for different salinity levels is shown in Fig. 18, for salt water solutions with different weight concentrations ranging from 0% to 19%. It can be observed that the resonance wavelength moves towards the shorter wavelength region (shift of ≈ 2 nm for the salt concentration range tested). This is due to the fact that, with increasing concentration of salt, there is an increase in the refractive index of the solution. Also, it turns out that the concentration of the salt solution does not substantially affect the amplitude of the LPG loss band (variation less than 0.07 dB). The resonance wavelength shift with the salt concentration, from where it can be derived a sensitivity of ≈ -0.10 nm/% Salt.

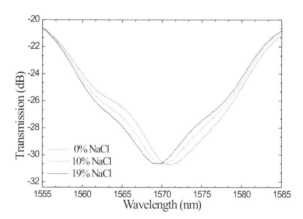

Fig. 18. Transmission spectra of the integrated LPG sensing head in an aqueous solution containing increasing concentration of salt.

4.3 Temperature measurement

To study the influence of the temperature on the developed structure, the system was placed into a thermostatic oven. The temperature was changed in increments of 5°C from 25° C to 100° C, in the air. The results are shown in Fig. 19. It is evident that the wavelength resonance shift is approximately linear up to ≈40°C, with a slope of ≈46 pm/°C. At this temperature there is a change, certainly due to some thermal induced mechanical adjustment in the sensing structure. From 45°C to 100°C, the behaviour keeps approximately linear but now with a slope of ≈ 60 pm/°C. Anyway, the sensing head was designed for application in aqueous environments where the expected temperature range is between 0 °C and 30°C. For this range, the thermal sensing behaviour is fairly linear and reversible, with a coefficient (≈ 40 pm/°C), which is different from the one found in a free LPG written in standard SMF28 fibre (≈ 70 pm/°C). This difference is certainly due to the fact that the LPG in the sensing head is under tension, and therefore the thermal induced shift of the LPG resonance is also affected by the thermal behaviour of the sensing head mechanical structure.

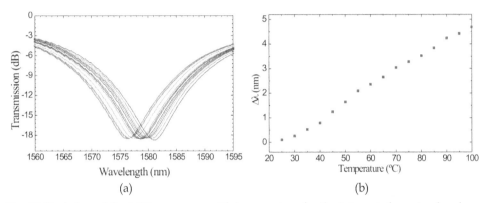

(a)   (b)

Fig. 19. Variation of the LPG resonance with temperature for the integrated sensing head.

From the results obtained it is clear that the problem of discriminating temperature induced shifts from refractive index induced wavelength shifts must be addressed in a real operational system. Nevertheless, many solutions are described in the literature that can be directly applied in our sensing head. The inscription of a single FBG in the vicinity of the LPG, for instance, would allow to univocally retrieving the temperature. Other option, was recently described [Jesus et al., 2009], using a pair of FBG that were inscribed after the LPG. However, their spectral position was chosen in such a way that each FBG was in opposite sides of the LPG resonance peak. Therefore, in this situation, the ratio of the reflected intensities of each FBG was modulated by the LPG shift being responsive to temperature and refractive index. On the other hand, the FBG wavelength was responsive only to temperature. The authors demonstrated that such a system can be operated in reflection, doubling the sensitivity to refractive index and enabling simultaneous determination of temperature and refractive index. From the practical point of view, such a system could be easily adapted into our robust sensing head protection system.

5. Conclusion

The preservation of species found in estuaries and coastal environments is achieved through constant monitoring of water quality, particularly parameters such as temperature and salinity. In the field of environmental monitoring, optical fibre refractometers (OFR) offer many advantages over conventional methods, such as immunity to electromagnetic interference, low weight, small dimensions and low insertion losses. Because of these favourable characteristics, the interest in developing fibre optic as salinity sensors has increased in the past few years, some of them based on refractive index measurements. In particular, long-period fibre gratings (LPGs) have attracted a lot of attention because of their high sensitivity to the refractive index of the surrounding environment.

This chapter focused the possible application of long-period gratings technology in environmental monitoring particularly in the measurement of surrounding refractive index or salinity. The features of this grating type were presented and it was proposed and experimentally demonstrated a prototype of a salinity sensor based on a LPG. The sensor element was written on a standard single-mode telecommunication optical fibre (Corning SMF-28) by electric arc discharge. By reducing the cladding diameter of the sensor, an improved measured resolution of 5.75×10^{-5} RIU over a measureable refractive index range from 1.33 to 1.43 (at the wavelength of 1550 nm) can be achieved, for a cladding diameter of 81 µm. It was analyzed the response of different gratings to refractive index changes with and without the structure developed, as well as its thermal response. Some important issues related with the LPG implementation in real conditions were presented, regarding the mechanical requirements and a filtration system of a housing support. The proposed robust sensing head has a filtration system for organic materials using the process of membrane separation. The developed system is easy to implement and allows the replacement of the filter without interfering with the rest of the process. Its performance analysis allowed to further underline the engineering aspects that must be taken in consideration when projecting such a system.

6. Acknowledgment

The authors gratefully acknowledge José Luis Santos and Orlando Frazão for their advice and crucial contribution. A special thanks to João Martins that designed the mechanical project for the sensing head. The acknowledgments also go to Pedro Corte-Real for his contribution with the schematics presented in this chapter. Finally, we would like to thank to José Manuel Rebordão for his encouragement.

7. References

Akiyama, M.; Nishide, K.; Shima, K., Wada, A. & Yamauchi, R. (1998). A novel long-period fibre grating using periodically releases residual stress of pure-silica core fiber. Optical Fiber Communication Conference and Exhibit, Technical Digest, pp. 276-277, ISBN 1-55752-521-8, San Jose, CA , USA, pp. 22-27.

Allsop, T.; Floreani, F.; Jedrzejewski, K.; Marques, P. & Romero, R. (2006). Spectral characteristics of tapered LPG device as a sensing element for Refractive Index and Temperature. Journal of Lightwave Technology., Vol. 24, No.2, pp. 870-878.

Bhatia, V. & Vengsarkar, A. (1996). Optical fiber long-period grating sensors. Optics Letters, Vol.21, No.9, pp. 692-694.

Bhatia, V. (1999). Applications of long-period gratings to single and multi-parameter sensing. Optics Express, Vol.4, No.11, pp. 457-466.

Cardenas-Sevilla, G.; Monzon-Hernandez, D.; Torres-Gomez, I. & Martinez-Rios, A., (2009). Mechanically induced long-period fiber gratings on tapered fibers. *Optics Communications*, Vol.282, No.15, pp. 2823-2826.

Cole-Parmer catalogue (2005/06) Cole-Parmer Instrument Company, US.

Davis, D.; Gaylord,T.; Glytis, E.; & Mettler, S. (1998).Long period fibre grating fabrication with focused CO_2 laser pulses. *Electronics Letter*, Vol.34, No.3, pp. 302-303, ISSN 0013-5194.

Ding, J.; Zhang, A.; Shao, L.; Yan J. & Sailing H. (2005). Fiber-taper seeded long-period grating air as a highly sensitive refractive-index sensor. IEEE *Photonics Technology Letters*, Vol.17, No.6, pp. 1247-1249.

Dong, X., L. Su, P. Shum, Y. Chung, & C. C. Chan (2005). Photonic-Crystal-Fiber-Based Mach–Zehnder Interferometer using Long-Period Gratings. *Optics Communications*, Vol.258, No.159, pp.1379-1383.

Dong, X.; Feng, S. ; Xu, O.; Lu, S.; & Pei, L. (2009). Add/drop channel filter based on two parallel long-period fiber gratings coupler. *Optik*, Vol. 120, pp. 855-859.

Erdogan, T. (1997a). Fiber grating spectra. *Journal of Lightwave Technology*. Vol.15, No.8, pp. 1277-1294.

Erdogan, T. (1997b). Cladding-mode resonances in short- and long-period fiber grating filters. *Journal of the Optical Society of America A*, Vol.14, pp.1760-1773.

Falate, R.; Frazão, O.; Rego, G.; Fabris, J. & Santos, J. (2006). Refractometric sensor based on a phase-shifted long-period fiber grating. *Applied Optics*, Vol.45, No.21, pp. 5066-5072.

Falciai, R., Mignani, A., & Vannini, A. (2001).Long period gratings as solution concentration sensors Sensors and Actuators B: Chemical, Vol.74, No.1-3, 15 pp. 74-77. Proceedings of the 5th European Conference on Optical Chemical Sensors and Biosensors.

Fujimaki, M.; Nishihara, Y., Ohki, Y. ; Brebner, J. & Roorda S. (2000). Ion-implantation-induced densification in silica-based glass for fabrication of optical fiber gratings. *Journal of Applied Physics*, Vol.88, No.10, pp. 5534-5537, ISSN 0021-8979.

Frazão, O.; Falate, R.; Fabris, J.; Santos, J. & Ferreira, L. (2006). Optical inclinometer based on a single long-period fiber grating combined with a fused taper. *Optics Letters*, Vol. 31, No. 20, pp 2960-2962.

Gwandu B., Shu X., Allsop T., Zhang W. & Zhang L. (1999). Simultaneous refractive index and temperature measurement using cascaded long period grating in double cladding fibre. *Electronics Letters*, Vol.35, No.14, pp. 695 – 696, ISSN 0013-5194.

Han, Y.; Lee, B.; Han, W.; Paek, U. & Chung, Y (2001). Fiber optic sensing applicatios of a pair of long period fibre gratings. *Measurement Science and Technology*, Vol.12, 778-781.

Hill, O.; Malo, B.; Vineberg, K., Bilodeau, F.; Johnson, D. (1990). Efficient mode conversion in telecommunication fiber using externally written gratings. *Electronics Letters*, Vol. 26, pp. 1270-1272.

Humbert G. & A. Malki (2002). Characterizations at very high temperature of electric arc-induced long-period fiber gratings. *Optics Communications*, Vol.208, No.4-6, pp. 329-335.

Humbert, G. & Malki A. (2003). High performance bandpass filters based on electric arc-induced ɪɪ-shifted long-period fibre gratings. *Optics Express*, Vol.39, No.21, pp. 1506-1505, ISSN 0013-5194.

James, S.; Rees, N.; Ashwell, G. & Tatam, R. (2002). Optical Fiber long-period Grating with a Langmuir –Blodgett thin film overlays. *Optics Letters*, Vol.9, pp. 686-688.

James, S. & Tatam, R. (2003). Optical fibre long-period grating sensors: Characteristics and application. *Measurement Science and Technology*, Vol. 14, pp. R49-R61.

Jesus, C.; Caldas, P.; Frazão, O.; Santos, J.; Jorge, P. & Baptista, J. (2009).Simultaneous Measurement of Refractive Index and Temperature Using a Hybrid FBG/LPG Configuration. *Fiber Integrated Optics*, Vol.28, pp 440–449.

Kashyp, R. (2010). Fiber Bragg Gratings. 2nd ed, Optics and Photonics Series, Academic Press, Boston.

Khaliq, S.; James, S.; & Tatam R. (2002). Enhanced sensitivity fibre optic long period grating temperature sensor. *Measurement Science and Technology*, Vol.13, No.5, pp 792–795.

Kim, B. ; Ahn, T. ; Kim, D. ; Lee, B. & Chung, Y. (2002). Effect of CO2 laser irradiation on the refractive-index change in optical fibers. *Applied Optics*, Vol.41, No.19, pp. 3809-3815.

Kondo, Y.; Nouchi, K.; Mitsuyu, T.; Watanabe, M. & Kazansky, P. (1999). Fabrication of long-period fiber gratings by focused irradiation of infrared femtosecond laser pulses, *Optics Letters*, Vol. 24, pp. 646-648.

Lee, B.; Liu, Y.; Lee, S.; Choi, S. & Jang, J. (1997). Displacement of the resonant peaks of a long-period fiber grating induced by a change of ambient refractive index. *Optics Letters*, Vol.22, pp. 1769-1771.

Lee, B.; Nishii, J. (1998). Self-interference of long period fibre gratings and its application as temperature sensor. *Electronics Letters*, Vol.34, pp. 2059-2060.

Lemaire, P.; Atkins, R.; Mizrahi, V. & Reed W. (1993). High pressure H2 loading as a technique for achieving ultrahigh UV photosensitivity and thermal sensitivity in GeO$_2$ doped optical fibres. *Optics Express*, Vol.29, pp. 1191.

Liu, Y.; Tu, W.; Yang, D. & Wang T. (2011), Fabrication of long-period fiber gratings by CO$_2$ laser in fiber under tension. *Journal of Shanghai University*, (English Edition) , Vol.15, No.1, pp. 1–6.

Murphy, R.;   James, S. &   Tatam, R. (2007). Multiplexing of Fiber-Optic Long-Period Grating-Based Interferometric Sensors. Journal of Lightwave Technology, Vol.25, No.3, pp. 825–829.

Mishra, V.; Singh, N.; Jain, S.; Kaur, P. & Luthra, R. (2005). Refractive index and concentration sensing of solutions using mechanically induced long period grating pair. *Optical Engineering*, Vol. 44, pp. 094402-1–094402-4.

Nikogosyan D. (2006). Long-period Gratings in a standard telecom fibre fabricated by high-intensity fentosecond UV and near-UV laser pulses. *Measurement Science and Technology*. Vol.17, pp. 960-967.

Patrick, H.; Chang, C. & Vohra, S. (1998a). Long period fiber gratings for structural bending sensing. *Electronics Letters*, Vol.34, pp. 1773 – 1775.

Patrick, H., Kersey, A. & Bucholtz, F. (1998b). Analysis of the response of long-period fiber gratings to external index of refraction. *Journal of Lightwave Technology.* Vol.16, pp. 1606–1612.

Queiroz S.; Collins, C. &; Jardim, I. (2001) .Métodos de extracção e/ou concentração de compostos encontrados em fluidos biológicos para posterior determinação cromatográfica. Quim. Nova, 24, N° 1, 68-76.

Rego G., O. Okhotnikov, E. Dianov, & V. Sulimov, (2001). High-temperature stability of long period fiber gratings produced by using an electric-arc. *Journal of Lightwave Technology*, Vol.19, pp. 1574-1579.

Rego G., Fernandes,J.; Santos, J.; Salgado, H. & Marques, P. (2003). New technique to mechanically induce long-period fibre gratings. *Optics Communications*, Vol. 220, No. 1-3, pp. 111-118.

Rego, G. ; Marques, P. ; Santos, J. & Salgado, H. (2005). Arc-induced Long-period Gratings. *Fiber and Integrated Optics*, Vol.24 (3-4), pp. 245-259.

Rego, G.; Santos, J.; & Salgado, H. (2006). Refractive index measurement with long-period gratings arc-induced in pure-silica-core fibres. *Optics Communications*, Vol.259, pp. 598–602.

Shifler D. (2005). Understanding material interactions in marine environments to promote extended structural life. *Corrosion Science*, Vol.47, pp. 2335–2352.

Shigley J. (1986). Mechanical Engineering Design - First Metric Edition, McGraw-Hill S Book Company, ISBN 0-07-056898-7.

Silva, C.; Coelho, J.; Caldas, P. ; Frazão, O.; Jorge, P.; Santos, J. (2010). Optical fibre sensing system based on long-period gratings for remote refractive index measurement in aqueous environments. *Fiber and Integrated Optics*, Vol.29, No.3, 160 — 169.

Swart P. L. (2004). A single fiber Michelson interferometric sensor. Proceedings of the 16 th International Conference on Optical Fiber Sensors- Japan, pp. 602-605.

Vengsarkar A.; Lemaire, P.; Judkins, J.; Bhatia, V. & Sipe, J.(1996a). Long-Period Fiber-Grating-Based Gain Equalizers. *Optics Letters*, Vol. 21, pp. 335-338.

Vengsarkar A., Pedrazzani, J.; Judkins J. B.; Lemaire, P. & Bhatia, V. (1996b). Long-period fiber gratings as band-rejection filters. *Journal of Lightwave Technology*, Vol.14, No.1, pp. 58-65.

Yokouchi T., Suzaki,Y.; Nakagawa, K.; Yamauchi, M. & Kimura (2005). Thermal tuning of mechanically induced long-period fiber grating. *Applied Optics*, Vol.44, No.24, pp. 5024-5028.

Yu, X.; Shum, P. & Dong, X. (2006). Photonic-crystal-fiber-based-Mach-Zehnder interferometer using long-period gratings. *Microwave and optical technology letters*, Vol.48, No.7, pp. 1379-1383.

Zhang, A.; Shao, L.; Ding, J. & He, S. (2005). Sandwiched long-period gratings for simultaneous measurement of refractive index and temperature. IEEE *Photonics Technology Letters*, Vol.7, No.11, pp. 2397–2399, ISSN 1041-1135.

Zanlorensi R. V., Costa, R. C. Kamikawachi, M. Muller and J. L. Fabris, (2009). Thermal characteristics of long-period gratings 266 nm UV-point-by-point induced. *Optics Communications*, Vol.282, No.5, pp 816-823.

Zhao Y., Y. Liao, B. Zhang, (2003). Monitoring Technology of Salinity in Water With Optical Fiber Sensor. *Journal of Lightwave Technology*, Vol.21, No.5, pp. 1334-1338.

Zhu Y., J. H. Chong, M. K. Rao, H. Haryono, A. Yohana, Ping Shum, Lu, C. (2003).A long-period grating refractometer: measurements of refractive index sensitivity. Proceedings SBMOIIEEE MTT-S IMOC, pp. 901-904.

Fiber Optic Displacement Sensors and Their Applications

S. W. Harun[1,2], M. Yasin[1,3], H. Z. Yang[1] and H. Ahmad[1]
[1]Photonic Research Center, University of Malaya, Kuala Lumpur
[2]Department of Electrical Engineering, Faculty of Engineering,
University of Malaya, Kuala Lumpur
[3]Department of Physics, Faculty of Science and Technology,
Airlangga University, Surabaya
[1,2]Malaysia
[3]Indonesia

1. Introduction

Optical fiber-based sensor technology offers the possibility of developing a variety of physical sensors for a wide range of physical parameters (Nalwa, 2004). Compared to conventional transducers, optical fiber sensors show very high performances in their response to many physical parameters such as displacement, pressure, temperature and electric field. Recently, high precision fiber displacement sensors have received significant attention for applications ranging from industrial to medical fields that include reverse engineering and micro-assembly (Laurence et al., 1998; Shimamoto & Tanaka, 2001); Spooncer et al., 1992; Murphy et al., 1991). This is attributed to their inherent advantages such as simplicity, small size, mobility, wide frequency capability, extremely low detection limit and non-contact properties. One of the interesting and important methods of displacement measurement is based on interferometer technique (Bergamin et al., 1993). However, this technique is quite complicated although it can provide very good sensitivity. Alternatively, an intensity modulation technique can be used in conjunction with a multimode fiber as the probe. The multimode fiber probes are preferred because of their coupling ability, large core radius and high numerical aperture, which allow the probe to receive a significant amount of the reflected or transmitted light from a target (Yasin et al., 2009; Yasin et al., 2010; Murphy et al., 1994). For future applications, there is a need for better resolution, longer range, better linearity, simple construction and low cost unit.

In this chapter, fiber-optic displacement sensors (FODS) are demonstrated using an intensity modulation technique. This technique is one of the simplest techniques for the displacement measurement, which is based on comparing the transmitted light intensity against that of the launch light to provide information on the displacement between the probe and the target. A silicon photo-diode is used to measure the transmitted and reflected light intensity. The sensor performances are investigated for various laser sources, different probes types and arrangements and different target. The theoretical analysis and the corresponding results on various bundled fiber based sensors are also presented in this chapter. The application of the FODSs in liquid refractive index measurement is investigated theoretically and experimentally. In the last part of this chapter, a continue monitoring the liquid level is also demonstrated by using the FODS.

2. FODS with beam-through technique

The intensity based sensors can be achieved by either beam-though or reflective techniques. A change in displacement of the through-beam and reflective sensors are manifested as a variation in the transmitted light and reflected light intensity, respectively. This section demonstrates a simple design for an intensity-based displacement sensor using a multimode plastic fiber in conjunction of beam-through technique. The performances of this sensor are investigated for both lateral and axial displacements. In the sensor, light is transmitted through a transmitting fiber to a receiving fiber and the received light is then measured by a silicon detector.

Fig. 1 shows a schematic diagram of the proposed sensor, which consists of two set of fiber, one set is connected to a light source and is termed as the transmitting fiber, and the other set is connected to a silicon detector and is known as the receiving fiber. In the experiment, the transmitting fiber located opposite to the receiving fiber is moved laterally and axially. The light is scattered after travelling out from the transmitting fiber and the receiving fiber collect a portion of the scattered light to transmit into the silicon detector where its intensity is measured. The intensity of the collected light is a function of axial and lateral displacement of the fiber. The light source is a He-Ne laser with a peak wavelength of 633 nm. The light is modulated externally by chopper with a frequency of 200 Hz, which is connected to lock-in amplifier to reduce the dc drift and interference of ambient stray light. For an axial and lateral displacement, a flexible adjusting mechanism

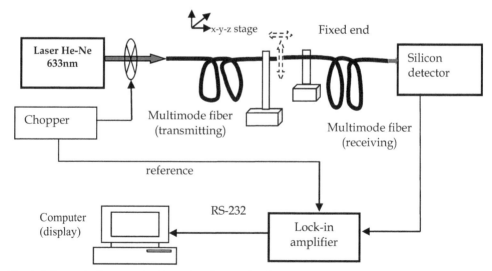

Fig. 1. Schematic diagram for lateral and axial displacement sensing using beam-through technique.

using piezoelectric is required, so the receiving fiber tip is mounted on a translational stage, which provides fine movement of the transmitting fiber surface in the axial and lateral direction. In this experiment, the axial and lateral micro distance is varied and the lock-in amplifier output voltage of the transmitted light is directly recorded by a computer automatically using Delphi software through serial port RS232. The piezoelectric micrometer can provide precise changes of about 25 and 30 nm for every positive and negative pulse, respectively, and in this experiment the displacement measurement is taken in successive steps of 45nm.

To analyze the performance of the proposed FODS, the output voltage from a receiving fiber is related to the axial and lateral displacements of the transmitting fiber. Both fibers should be mounted perpendicular to each other and positioned flush against the surface. The output voltage of the sensor should be highest in this position. The transmitting fiber is then moved away laterally and axially from the receiving fiber tip by still maintaining perpendicularity between them. Fig. 2 shows the output voltage of the lock-in amplifier against the lateral displacement between the two ends of the fibers. In the experiment, the core diameter for both ends is varied. As expected, the voltage is highest at zero displacement from the center and the lateral movements of the transmitting fiber away from the receiving fiber resulted in a reduced output voltage as shown in the figure. The power drop pattern follows the theoretical analysis by Van Etten & Van der Plaats (1991) in which the output transmission function is given by:

$$\eta = \frac{2}{\pi}\left(\left(acr\cos\left(\frac{d}{a}\right) - \frac{d}{a}\sqrt{1-\left(\frac{d}{a}\right)^2}\right)\right) \tag{1}$$

where η, d, and a are coupling efficiency, lateral displacement, and fiber core radius, respectively. η is defined as the ratio of output voltage over the maximum voltage. The sensitivity of the sensor is determined by a slope of a straight-line portion in the curves.

As shown in Fig. 2, the beam-through type of sensor has two symmetrical slopes and the sensitivity is higher at the smaller core diameter. At core diameter of 0.5 mm for both transmitting and receiving fibers, the sensitivity is obtained at around 0.0008mV/μm and the slope shows a good linearity of more than 99% within a range of 420μm. The linear range increases to around 800μm for the both slopes as the core diameter increased to 1.0 mm. The linear range can be further increased to more than 1000μm by using a larger core for the receiving fiber as shown in Fig. 2. However, the voltage is unchanged at a small lateral displacement for this sensor due to the larger receiving core, which covers the whole diverged beam from the transmitting fiber. The highest resolution of 13μm is obtained with core diameter of 0.5 mm for both fibers. In this work, the resolution is defined as the minimum displacement which can be detected by this sensor.

The performance of the sensor with lateral displacement is summarized as shown in Table 1. The output voltage from a receiving fiber is related to axial displacement of end surface of the transmitting fiber. Fig. 3 shows the output voltage of the lock-in amplifier against the axial displacement for the different core diameters. In this experiment, the end surface of transmitting fiber is moved away from the receiving fiber tip by still maintaining perpendicularity between them. As expected, the voltage is highest at zero displacement from the center and the output voltage reduces as the axial displacement increases for all

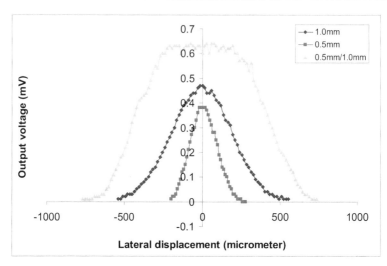

Fig. 2. The output voltage of the lock-in amplifier against the lateral displacement of the transmitting fiber.

Fiber's core diameter	The left slope		The right slope		Resolution (µm)
	Sensitivity (mV/µm)	Linear range (µm)	Sensitivity (mV/µm)	Linear range (µm)	
1.0mm /1.0mm	0.0003	855(585-1440)	0.0004	765 (1845-2610)	33
0.5mm /0.5mm	0.0008	420(120-540)	0.0007	420 (750-1170)	13
0.5mm /1.0mm	0.0005	1035(450-1485)	0.0005	1125 (3060-4185)	20

Table 1. Performance of the Lateral Displacement Sensor

core diameters used. The power drop pattern follows the theoretical analysis from (Van Etten & Van der Plaats, 1991) in which the output transmission function is given by:

$$\eta \approx 1 - \frac{z}{a} \frac{2}{\pi (NA)^2} \left(\arcsin(NA) - NA \sqrt{1 - (NA)^2} \right) \qquad (2)$$

where η, z, a, and NA are coupling efficiency, axial displacement, core radius, and numerical aperture, respectively.

As shown in Fig. 3, the sensors only have one slope and the sensitivity is higher at the smaller core diameter. At core diameter of 0.5 mm (for both transmitting and receiving fibers), the sensitivity is obtained at around 0.0002 mV/µm, which is the highest and the slope shows a good linearity of more than 99% within a range of 900µm. The linear range increases to 3195 m with the larger core diameter of 1.0 mm. In case of the receiving core is bigger than the transmitting core, the voltage is almost constant at small axial displacements due to the coherent light source, which has a small divergence angle. The highest resolution

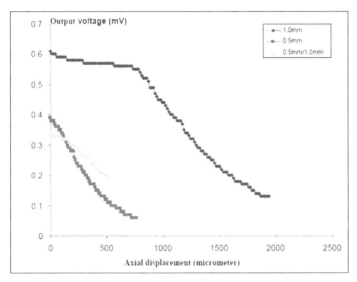

Fig. 3. The output voltage of the lock-in amplifier against the axial displacement of the transmitting fiber.

is obtained at 50 m with 0.5 mm core diameter for both fibers. The performance of the sensors with axial displacement is summarized as shown in Table 2. The stability of these sensors is observed to be less than 0.01 mV (3%). The experimental results are capable of offering quantitative guidance for the design and implementation of the displacement sensor. This sensor requires two probes, which is precisely aligned and therefore the applications are limited. In the next section, various reflective types of FODSs are proposed to improve the sensitivity, linearity and dynamic range of the sensor.

Fiber's core diameter	Sensitivity (mV/μm)	Linear range (μm)	Resolution (μm)
1.0mm/1.0mm	0.0001	3195 (2340-5535)	100
0.5mm/0.5mm	0.0002	900(0-900)	50
0.5mm/1.0mm	0.0001	1530(0-1530)	100

Table 2. Performance of the Axial Displacement Sensor

3. FODS with reflective configuration

The FODS based on intensity modulation and reflective arrangement provides a promising solution for displacement measurement in terms of wide dynamic range, with high potential for ultra-precise non-contact sensing. It also provides flexibility in incorporating the optical sensors permanently into composite structures for monitoring purpose [Wang et al., 1997). In the simplest design of reflective FODS, a probe with a pair of fibers is normally used as the media to transfer/collect the light to/from the target and its theoretical analysis is fully contributed (Faria, 1998). In the design of FODS system the sensor probe is playing a majority role comparison with the selection of laser source and reflector. Hence, the researchers are paid more attentions in the development of sensor probe to improve the performance of FODS. The FODSs described in this section are focus on the various configurations of sensor probe since they are mainly influence the performance of FODS.

3.1 FODS using a probe with two receiving fibers

In this section, a new configuration of the FODS is reported by using two receiving fibers which are bundled together. The mathematical analysis of FODS is developed to simulate the theoretical results, which is then compared to the experimental result. The performance of this FODS is also compared experimentally with the conventional FODS. The probe structure of the proposed FODS is shown in Fig. 4. It consists of one transmitting and two receiving plastic multi-mode fibers bundled together in parallel. To analyze the theory of this sensor setup, a more realistic approach – Gaussian beam is used to describe the light leaving the transmitting fiber. The irradiance of emitted light is obeying an exponential law according to

$$I(r,z) = \frac{2P_E}{\pi \omega^2(z)} \exp\left(-\frac{2r^2}{\omega^2(z)}\right) \tag{3}$$

where P_E is the emitted power from the light source, r is the radial coordinate and z is the longitudinal coordinate. $\omega(z)$ is the beam radius which is also a function of z, $\omega(z) = \omega_0\sqrt{1 + (\frac{z}{z_R})^2}$. The waist radius ω_0 and Rayleigh range z_R are the important parameters in the Gaussian Beam function.

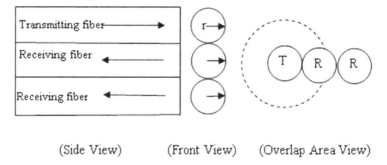

(Side View)   (Front View)   (Overlap Area View)

Fig. 4. Side, Front and Overlap Views of FODS probe structure with two receiving fibers

The optical power received by the receiving fiber can be evaluated by integrating the irradiance, I over the surface area of the receiving fiber end, S_r.

$$P(z) = \iint_{S_r} I(r,z)dS_r \tag{4}$$

The overlapping area of the reflected light area and the core of the receiving fibers is also illustrated in Fig 4. The power value of reflected light collected by two receiving fibers increased with the increased of displacement of probe and target mirror. Two receiving fibers collected the reflecting light results the receiving light power increased. Based on this geometrical analysis two receiving fibers collected the reflected light significantly affects the transfer function of the FODS.

The Eq. (3) can be described in other expressions in order to simulate conveniently. The power collected by the first and second receiving fibers are denoted as P_1 and P_2, respectively where P_1 is closer to the transmitting fiber. From Eq. (3), P_1 and P_2 can also be written as;

$$P_1(z) = \frac{2P_E}{\pi \omega^2(z)} \int_{y=-R_r}^{R_r} \int_{x=R_t+R_r-\sqrt{R_r^2-y^2}}^{R_t+R_r+\sqrt{R_r^2-y^2}} \exp\left(-\frac{2(x^2+y^2)}{\omega^2(z)}\right) dx dy \tag{5}$$

$$P_1(z) = \frac{2P_E}{\pi \omega^2(z)} \int_{y=-R_r}^{R_r} \int_{x=R_d+R_t+R_r-\sqrt{R_r^2-y^2}}^{R_d+R_t+R_r+\sqrt{R_r^2-y^2}} \exp\left(-\frac{2(x^2+y^2)}{\omega^2(z)}\right) dx dy \tag{6}$$

where the R_t, R_r is the radius of transmitting fiber and receiving fiber, respectively. The R_d is the diameter of the receiving fiber. The radial coordinate r is expressed by $\sqrt{x^2+y^2}$ in Cartesian coordinate system. Then the total amount of power collected by both receiving fibers is;

$$P = P_1 + P_2 \tag{7}$$

The conventional FODS only collects power P_1 using a pair of fibers bundled together. Compared to this sensor, the proposed FODS collects more reflected light due the additional P_2 power, which increases the dynamic range of the sensor.

The software simulation is programmed and implemented in MATLAB. Some important parameters are specified in the programming, wavelength of the laser source λ = 594nm, transmitting fiber core radius R_t = 0.5mm and numerical aperture value NA = 0.25. Fiber diameter R_d = 2mm. The theoretical analysis transfer function of proposed displacement sensor in Eq. (6) can be normalized by its maximum collected power P_{max}, $P_n = P/P_{max}$. The normalized distance $h_n = h/z$, while power increase to maximum for small values h from 0 to z, power decrease to zero according to the h values higher than z. The simulation results are then compared with the experimental results.

The experiment setup for the FODS with two receiving fibers is shown in Fig. 5. It consists of a light source, a chopper, a sensor probe, a flat mirror, a silicon detector, a lock-in amplifier and a PC. The sensor probe consists of one transmitting and two receiving plastic multi-mode fibers which are bundled together in parallel. The transmitting and receiving fiber length is 2m with a core diameter of 1mm. A 594nm yellow He-Ne laser is used as a light source. It has the maximum output power of 1mW and beam divergence of 0.92mRads, which is modulated by external chopper with a frequency 200Hz. The transmitting end of fiber probe radiates the modulated light from the laser to the mirror. The flat mirror is controlled by a piezoelectric motor and driver. The distance between the fiber probe and the mirror is varied in successive steps of 4µm and the light voltage which is represented the optical power is measured against the change in the mirror displacement stage. Then the mirror reflects the transmitting light into the receiving end of fiber probe. The reflected light through receiving end can be detected by the silicon detector. The photon energy collected by detector is converted into a voltage. The output of the detector transfers into the lock-in amplifier to deduce the dc drift and filter out the undesired noise. The lock-in amplifier is connected to a PC using RS232 data series line. From the PC, the output light voltage is monitored.

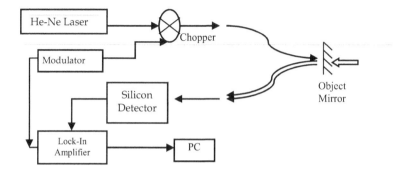

Fig. 5. Experiment setup of proposed FODS with two receiving fibers.

In both theoretical and experimental analysis, the results are processed and displayed in the normalized forms which the output power is normalized by the maximum collected optical power and the displacement is normalized by the distance parameter z. This is to make the output function a dimensionless function and eliminate the dependency of the FODS output function to the fiber core radius and divergence angle. Fig. 6 shows the comparison of the simulated and the observed response of the proposed FODS. As shown in Fig. 6, both curves exhibit a maximum with a steep linear front slope and back slope which follows an almost inverse square law relationship for the reflected light intensity versus distance of the mirror from transmitting fiber end. The signal is minimal at zero distance because the light cone does not reach the core of both receiving fibers. As the displacement increases, the size of cone of the reflected light at the plane of fiber also increases, which then starts to overlap with the receiving fiber cores leading to a small output voltage. Further increases in the displacement lead to larger overlapping which in turn results in an increase in the output voltage. However, after reaching the maximum value, the output voltage starts decreasing even though the displacement increases. This is due to the large increase in the size of the light cone and the power density decreases with the increase in the size of the cone of light.

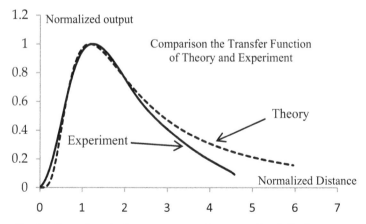

Fig. 6. Normalized collected optical power versus the normalized displacement curves for both theoretical and experimental results.

The maximum normalized output powers of 1 are obtained at the normalized displacement distances of 1.2 for the theoretical curve and 1.3 for the experimental curve as shown in Fig. 6. The close agreement of theoretical and experimental results in the Fig. 6 is quite evident. The small difference can be accounted for experimental limitations mainly due to the geometry error of the fiber used and positioning error. Table 3 summarizes the performance comparison of the simulated and the experimental result of the sensor. The slope of the response curve is the sensitivity, which is expressed in the unit of mW/μm. As shown in the table, the linearity range at the back slope is nearly 3 times that of the linearity range at the front slope. However, the sensitivity at the front slope is nearly 3~4 times higher than that of the sensitivity at the back slope.

Methods	The front slopes		The back slopes	
	Linearity range	Sensitivity	Linearity range	Sensitivity
Theoretical	0.8 μm (0.3 ~ 1.1)	0.06 mV/μm	2.5 μm (1.4 ~ 3.9)	0.015mV/μm
Experimental	0.9 μm (0.2 ~ 1.1)	0.06 mV/μm	2.5 μm (1.5 ~ 4.0)	0.017mV/μm

Table 3. The performance comparison between theoretical and experiment results.

The experiment is also repeated for the conventional sensor with one transmitting and one receiving fibers. The response of the conventional FODS is then compared to that of the proposed FODS as shown in Fig. 7. As shown in the figure, the maximum power collected by the receiver is obtained at a shorter distance for the conventional sensor. The sensor performance comparison between the proposed and the conventional sensors is summarized in Table 4. The sensitivity of both sensors is almost similar. However, the proposed sensor obviously has a better linearity range as shown in Table 4. This is attributed to the amount of the collected light intensity, which is higher in the proposed sensor compared to that of the conventional one.

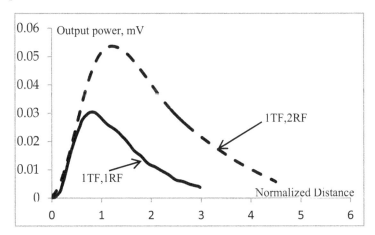

Fig. 7. The displacement curves for both proposed (1TF, 2RF) and conventional (1TF, 1RF) sensors

Methods	The front slopes		The back slopes	
	Linearity range	Sensitivity	Linearity range	Sensitivity
Conventional	0.5 μm (0.2 ~ 0.7)	0.06 mV/μm	1.4 μm (0.9 ~ 2.3)	0.016 mV/μm
Proposed	0.9 μm (0.2 ~ 1.1)	0.06 mV/μm	2.5 μm (1.5 ~ 4.0)	0.017 mV/μm

Table 4. The performance comparison between the proposed and conventional sensors

The maximum linearity ranges of 2.5 μm and 0.9 μm are obtained at the back and front slopes respectively for the proposed sensor as shown in Table 4. The linearity range of the proposed sensor is improved by about 44% for both slopes compared to the conventional sensor. As indicated by the above results, we can conclude that the employment of two receiving fibers increases the linearity range of the sensor, which is very useful for the large displacement measurements. Both theoretical and experimental results are capable of offering quantitative guidance for the design and implementation of the displacement sensor.

3.2 The FODS with two asymmetrical fibers bundled

As discussed in most of the FODSs [Ko et al., 1995; Elasar et al., 2002; Oiwa & Nishitani, 2004; Cao et al., 2007), the radius of the transmitting and receiving fibers are often made the same for the convenience of analytical study and experiments. However, there is a lack of research work on the displacement sensor using bundled fiber with different core radius. In this work, a mathematical model of displacement sensor using asymmetrical bundled fiber is developed. Some simulations were carried out based on the mathematical model and experimental results were also obtained to validate the MATLAB simulated results. The effect of different core radial ratios (CRRs) to the dynamic range, sensitivity and illumination area of bundled fiber are analyzed and discussed in this section.

Theoretical analysis

The proposed FODS consists of a transmitting and receiving fibers as well as a reflecting mirror. Both fibers are of different core radius and are bundled together in parallels as shown in Fig. 8. Let r_T and r_R denote the core radius of the transmitting fiber and the core radius of the receiving fiber. The core radial ratio, CRR is the ratio of transmitting fiber and receiving fiber core radius, as given below:

$$\text{CRR}, \ k = \frac{r_R}{r_T} \tag{8}$$

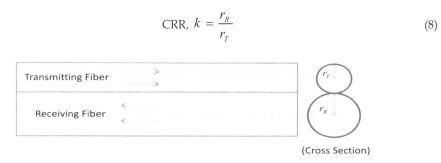

(Cross Section)

Fig. 8. Side and front views of the transmitting and receiving fiber ends.

Fig. 9 shows the geometrically illustration of the overlapping area between the reflected light circle and the core of the receiving fiber at different CRR. Based on this figure, in the same core radius of transmitting fiber; the reflected light power collected by the receiving fiber increases with the increasing core radius of the receiving fiber. The larger receiving fiber core radius and core area, the bigger fraction of reflected light can be collected by the receiving fiber. In the previous report (Faria, 1998), two major approaches have been introduced for theoretical analysis, namely geometrical and Gaussian Beam approaches. For the former approach, the simple assumption is made that the light intensity is constant within the reflected light circle. On the other hand, the light intensity outside the reflected light circle is null. This approach is apparently less accurate compared to the second approach. Gaussian Beam approach is a more realistic and more accurate method. The intensity of the light emitted from the transmitting fiber is described with Gaussian distribution as shown in Eq. (3).

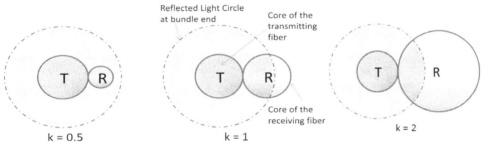

Fig. 9. Geometrical illustration for the overlapping area between the reflected light circle and the core of the receiving fiber at different CRRs.

The light power collected by the receiving fiber can be evaluated by using integral as shown in Eq. 4. However, the exact integration is tedious and impossible. Therefore, assumptions and approximation were used to dissolve the integration. For points situated in the far-field, $z >> z_R$ the following relations with the divergence angle can be obtained

$$\theta_a = \tan \theta_a = \frac{\omega(z)}{z} = \frac{\omega_0}{z_R} = \frac{\lambda}{\pi \omega_0} \tag{9}$$

Core radius of the transmitting fiber and receiving fibers are given by the approximation

$$\omega_T = z_a \tan \theta_a \approx z_a \theta_a \tag{10}$$

and

$$\omega_R = k \omega_T = k z_a \theta_a \tag{11}$$

where z_a is the distance between the beam source to the fiber end (Wang et al., 1997). The core area of the receiving fiber is computed from

$$S_a = \pi \omega_R^2 = \pi k^2 z_a^2 \theta_a^2 \tag{12}$$

The radial distance between the two core centers of the transmitting fiber and receiving fiber is determined from

$$r = \omega_T + \omega_R = \omega_T + k\omega_T = (1+k)z_a\theta_a \tag{13}$$

The path of the beam from the beam source to the bundle end after the reflection is given by

$$z_a + 2h \tag{14}$$

The displacement parameter in the normalized form is presented as

$$\zeta = \frac{z_a + 2h}{z_a} \tag{15}$$

or

$$\zeta = 1 + 2h_N \tag{16}$$

where $h_N = h/z_a$. To relate the displacement between the reflective mirror to the fiber end, h to the transfer function, with the help of the results determined above, the collected power of the receiving fiber can be expressed as

$$P(h) = \frac{2P_E}{\pi\omega^2(z_a + 2h)}\exp\left(-\frac{2r^2}{\omega^2(z_a + 2h)}\right) \times S_a$$

$$= \frac{2P_E k^2 z_a^2}{(z_a + 2h)^2}\exp\left(-\frac{2((1+k)z_a)^2}{(z_a + 2h)^2}\right) \tag{17}$$

By substituting Eq. (15) into this equation, we obtain

$$P(\xi) = \frac{2P_E k^2}{\xi^2}\exp\left(-\frac{2(1+k)^2}{\xi^2}\right) \tag{18}$$

The maximum received power is achieved when $P'(\xi) = 0$, and this leads to

$$\xi_{max} = \sqrt{2}(1+k) \tag{19}$$

Based on the relation in Eq. (3-16), the maximum h is given by

$$h_{max} = \frac{\sqrt{2}k + \sqrt{2} - 1}{2} \tag{20}$$

The maximum power is given by

$$P_{max} = P(\sqrt{2}(1+k)) = \frac{k^2 P_E}{(k+1)^2}\exp(-1) \tag{21}$$

In the normalized form, Eq. (3-15) is rewritten as

$$P_N(\xi) = \frac{P(\xi)}{P_{max}} = \frac{2(k+1)^2}{\xi^2} \exp\left(1 - \frac{2(1+k)^2}{\xi^2}\right) \tag{22}$$

In the analysis, the theoretical model of the FODS is modeled based on the similar parameters used in the experiment: Wavelength of the laser source λ = 594nm, transmitting fiber core radius r_R =0.5mm and numerical aperture value NA = 0.4. Based on the same parameters, four analytical model were simulated for k = 0.5, 1, 2 and 3 which were based on the available fiber core radius combinations in the experiments.

Experiment

The experiment for the FODS with two asymmetrical bundled fibers is carried using the similar set-up as shown in Fig. 5, but using bundled fiber with different core radius. The asymmetrical bundled fiber is constructed by pairing two different plastic fibers with the core radiuses of either 0.25mm or 0.50mm or 0.75mm. Due to the limited selections of core diameters, six combinations were selected for the experiments: $[k, r_T, r_R]$ = [0.5, 0.5mm, 0.25mm], [1.0, 0.5mm, 0.5mm], [2.0, 0.25mm, 0.5mm] and [3.0, 0.25mm, 0.75mm]. k is the core radial ratio. A precise displacement reference between the bundle end and the reflecting mirror is imperative for the experiment. Therefore, a New Focus 9061 motorized stages which is driven by a picomotor is used to change the displacement of the reflecting mirror from the fiber probe. Each increment step in the displacement is made identical and accurate. The collected light power in the receiving fiber is converted by a silicon detector into electrical power. Lastly, the electrical signal is filtered by a lock-in amplifier and recorded in the computer.

Characteristic of the sensor

In both theoretical and experimental analysis, the results are processed and displayed in the normalized forms which the output power is normalized by the maximum output power and the displacement is normalized by the parameter z_a. This is to make the output function a dimensionless function and eliminate the dependency of the FODS output function to the fiber core radius and divergence angle. Figs. 10 and 11 show the analytical and experimental results respectively for the proposed FODS. As shown in both figures, the location of the maximum output is shifted toward the right along the axis of displacement as the value of k increases. Besides, the linear range on the front slope and back slope gets larger for every larger value of k. Both graphs exhibit the almost the same trend of characteristics in the curves as the value of k becomes larger. This phenomenon can be explained by the use of distinctive core radius of the two fibers. As shown in Fig. 9, at the same displacement the fraction of overlap area in receiving fiber core by the reflected light circle (percentage of shaded area in the receiving fiber core) is differ for different CRR. For a larger value of k, the fiber displacement sensor requires further displacement to achieve maximum overlap area. Adversely, the sensitivity of the fiber displacement sensor decreases as the CRR increases. On the other hand, some error in the initial displacement (0< h < 0.3) is observed if the two overlaid graphs are compared. This error is accounted to the approximation used in the theoretical analysis.

The performance of the proposed FODS from the experimental results is summarized in Table 5. The results show that the magnitude of the sensitivity decreases as the CRR or k value becomes larger while the linear range is larger for a larger value of k. The sensitivity

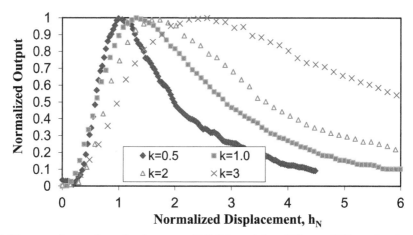

Fig. 10. The experimental result of proposed FODS model at different CRRs or k values.

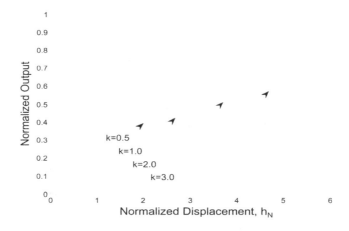

Fig. 11. The theoretical result of the fiber displacement sensor at different CRRs or k values.

CRR , k	Front Slope Sensitivity	Linear Range (mm)	Back Slope Sensitivity	Linear Range (mm)
0.5	1.7615	0.320 - 0.800	-0.5270	1.312 - 1.176
1	1.0984	0.230 - 0.998	-0.3378	1.536 - 3.302
2	0.9688	0.288 - 1.296	-0.2581	1.728 - 4.320
3	0.6955	0.320 - 1.440	-0.1366	2.240 - 5.400

Table 5. The sensitivity and linear range for different k values

characteristic trend is consistent with the theoretical plot as shown in Fig. 12. Fig. 12 shows the normalized sensitivity against normalized displacement at various k values. The curve width of the graph represents the linear range of the sensor. As shown in the figure, the linear range of the sensor increases with the value of k which is in agreement with the

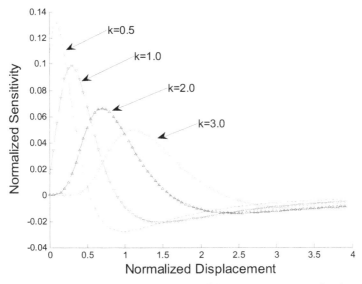

Fig. 12. Theoretical normalized sensitivity curves of the FODS at various k values.

experimental result in Table 5. This property provides a greater enhancement in FODS applications in terms of flexibility, wider dynamic range and high precision displacement measurement. The maximum sensitivity of 1.76 is obtained at k=0.5. The largest dynamic range of Fig. 10 is obtained at k=3. The conventional FODS which uses two fibers with identical core radius often encounters several restrictions due to limited linear range for the measurement. Besides, the limited option of sensitivity often becomes a challenge in the high-precision measurement. This restriction can be avoided using a suitable CRR or k value. The k value can be chosen in a way to provide the optimum performance.

3.3 The FODSs with two asymmetrical fibers inclined

To date, many works have been reported on the intensity modulation based FODSs (Bock et al., 2001; Cui et al., 2008; Saunders & Scully, 2007; Miclos & Zisu, 2001; Kulchin et al., 2007), which the probe consists of a pair of fibers used for transmitting and receiving the light. For instance, Buchade and Shaligram, et al. (2006) presented a FODS using two fibers inclined with a same angular angle and reported the sensitivity was enhanced compared with the conventional sensor with parallel bundled fibers. It also reported that the performances of the FODS with two fibers are depended mainly on four parameters: the offset, the lateral separation and the angle between the transmitting and receiving fiber tips, and the angle of the reflector (Buchade & Shaligram, 2007). However, there is still a lack of research work on the FODSs with different geometry of the receiving fiber. Hence, a study of this type of FODS is depicted in this section theoretically and experimentally.

Theoretical

Fig. 13 shows the geometry of the inclined displacement sensor, which consists of a transmitting fiber, receiving fiber and a reflector. The sensor performance is studied at various core radiuses of transmitting and receiving fiber. Fig. 13 (a) (Fig. 13(b)) shows the geometry of the sensor in case of the receiving core is bigger (smaller) than the transmitting core. Two asymmetrical transmitting and receiving fibers are mounted at an angle 'θ' with reference to the normal to the reflector. This ensures the receiving fiber core to collect the maximum power from cone of emitting light of the transmitting fiber. The shortest distances between the sensor probe tips and reflector are x_1 and x_2 for transmitting fiber and receiving fiber, respectively. The dash lines in receiving fiber is represented the size of diameter value same as the transmitting fiber. The image of transmitting (receiving) fiber is formed at a further distance x_1 (x_2) opposite to the transmitting (receiving) fiber beyond the reflector. The image fiber is thus seen located at $2x_1$ or $2x_2$ from the original position of the probe. Effectively the reflected light appears to form a cone and reaches the receiving fiber, which is parallel aligned in the cone as shown in Fig. 13.

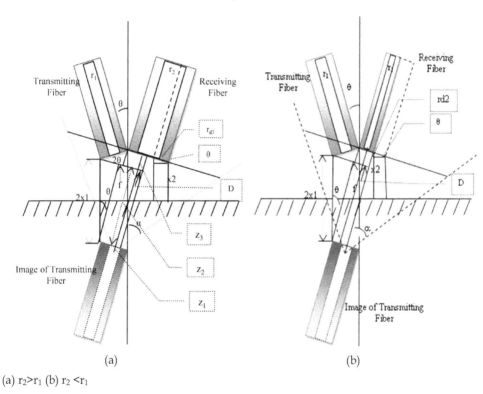

(a)   (b)

(a) $r_2 > r_1$ (b) $r_2 < r_1$

Fig. 13. The structure of sensor probes with two asymmetrical inclined fibers

As shown in Fig. 13, the core radius of the transmitting and receiving fibers are denoted as r_1 and r_2, respectively. Meanwhile, the diameters of transmitting fiber and receiving fiber are r_{d1} and r_{d2}, respectively. We assume that the ratio between the radius of two fibers is k_1, $k_1 = r_1/r_2$ and the ratio of diameter of two fibers is $k_2 = r_{d1}/r_{d2}$. From the geometry analysis of Fig. 13, the distance between the two sides of image of transmitting and receiving fibers is given by:

$$f = \left| r_{d1} \times \cos(2\theta) - 2x_1 \times \sin(\theta) \right| \tag{23}$$

Then, the distance between two fiber core, D is obtained as;

$$D = f + \frac{r_{d2} - r_{d1}}{2} = f + \frac{r_{d2}(1 - k_2)}{2} \tag{24}$$

The acceptance angle of transmitting fiber α is given by $\alpha = \sin^{-1}(\frac{NA}{n})$, where NA is a numerical aperture for the transmitting fiber. The distance between emitting point of transmitted light to the receiving flat core is denoted as z, which is given by;

$$z = z_1 + z_2 + z_3 \tag{25}$$

where $z_1 = r_1 \times cota = k_1 r_2 \times cota$, $z_2 = 2x_1 \times cos(\theta)$ and $z_3 = r_{d1} \times sin(2\theta) = k_2 r_{d2} \times sin(2\theta)$ as illustrated in Fig. 13.

To analyze the power collected by the receiving fiber, we simply analysis the light inside the fiber by using a Gaussion beam approach. The irradiance of emitted light is obeying an exponential law according to Eq. (3). The optical power received by the receiving fiber can be evaluated by integrating the irradiance, I over the surface area of the receiving fiber end which is given by Eq. (4). To simulate conveniently, the Eq. (4) can be described in other expressions;

$$P_1(k_1, k_2, z) = \frac{2P_E}{\pi \omega^2(z)} \int_{y=-r_2}^{r_2} \int_{x=D-\sqrt{r_2^2-y^2}}^{D+\sqrt{r_2^2-y^2}} \exp\left(-\frac{2(x^2 + y^2)}{\omega^2(z)}\right) dxdy \tag{26}$$

The $P(k_1, k_2, z)$ is the power collected by the receiving fiber corresponding the parameters k_1 and k_2. The radial coordinate r is expressed by $\sqrt{x^2 + y^2}$ in Cartesian coordinate system.

Fig. 14 illustrates the overlap area of the reflected light area and the core of the receiving fiber. The overlap area is zero at $x_2 = 0$ (Fig. 13(a)) or $x_1 = 0$ (Fig. 13(b)) and at a very small displacement (blind area) where the jacket of the two fibers blocks the reflected. As the displacement is increased further, the overlap area increases and thus increases the total power collected by the receiving core. The total power is maxima when the receiving cone covers the entire receiving core area. After that, the received optical power starts to decay exponentially as the displacement continues to increase. The received optical power is strongly dependent on the core size of the receiving fiber. At inclination angle of 2θ between the transmitting and receiving fibers, the distance x_1 between the sensor probe tip and reflector is given by (Buchade & Shaligram, 2006)

$$x_1 = rd_1(cosec\,\theta - 2sin\,\theta)/2 \tag{27}$$

From the geometrical analysis of Fig. 13, the distance x_2 is obtained as; $x_2 = x_1 - r_{d3}sin\theta$ for $r_{d1} < r_{d2}$ (Fig. 13(a)) or $x_2 = x_1 + r_{d3}sin\theta$ for $r_{d1} \geq r_{d2}$ (Fig. 13(b)) where $r_{d3} = r_{d2} - r_{d1}$. Therefore, the distance between sensor probe tip and reflector mirror can be summarized as;

$$x_2 = \frac{r_{d2}}{2}\left(k_2 \cos ec\theta + 2\sin\theta(1 - 2k_2)\right) \qquad (rd1 > rd2)$$

$$= \frac{k_2 r_{d2}}{2} \left(\cos ec\theta - 2\sin\theta \right) \qquad (rd1 = rd2)$$

$$= \frac{r_{d2}}{2} \left(k_2 \cos ec\theta - 2\sin\theta \right) \qquad (rd1 < rd2) \qquad (28)$$

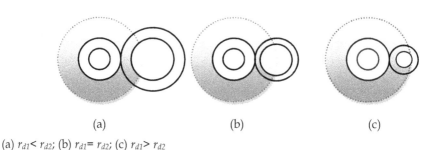

(a)   (b)   (c)

(a) $r_{d1} < r_{d2}$; (b) $r_{d1} = r_{d2}$; (c) $r_{d1} > r_{d2}$

Fig. 14. Overlap area view

The proposed sensor is simulated by using a MATLAB programming. To simplify the analysis, the k_1 values of 0.5, 0.667, 1, 1.5 and 2 are used. The k_2 value is set based on the availability of the fiber in our laboratory. In this simulation, the k_2 values of 0.5, 1 and 2 are used. The wavelength of the laser source λ is set at 594nm. The numerical aperture values $NA_1 = 0.27$, $NA_2 = 0.32$ and $NA_3 = 0.4$ are used for the core radius of 0.25mm, 0.5mm and 0.75mm, respectively.

Experiment

To verify the simulated results the FODS is constructed by mounting the transmitting and receiving fibers on the plastic board at angle with reference to the normal of the reflector as shown in Fig. 15. Separate samples with various fiber diameters and core radius are prepared for angle = 10º, 20º and 30º. Light from 594 nm He-Ne laser is modulated by an external chopper at frequency of 200 Hz and launched into the transmitting fiber. The light has an average output power of 3.0 mW, beam diameter of 0.75mm and beam divergence of 0.92 mRads. The length of transmitting and receiving fiber length is approximately 2m. The transmitting fiber radiates the modulated light from the light source to the target mirror, the displacement of sensor probe tip between mirror is controlled by a piezoelectric & driver. The reflective light from target mirror, which is mounted in the bottom of tank, is collected by the receiving fiber whose carriers the light into the silicon detector. A lock-in amplifier is connected with the detector to deduce the dc drift. The initial experiment is carried out by varying the inclination angle between the fibers.

Results and discussion

Fig. 16 compares the experimental and theoretical plots of the normalized output collected against normalized displacement between probe and reflector with air medium in between. In this study, the ratios k_1, k_2, and angle θ are set at 0.667, 0.5, 10° respectively. As shown in the figure, the theoretical curve is in good agreement with the experimental curve, verifying

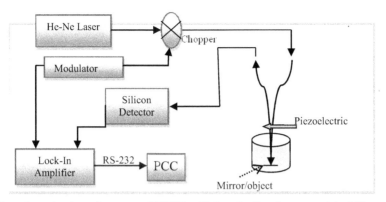

Fig. 15. Experiment setup of proposed FODS with two inclined asymmetrical fibers (Yang, et al., 2009).

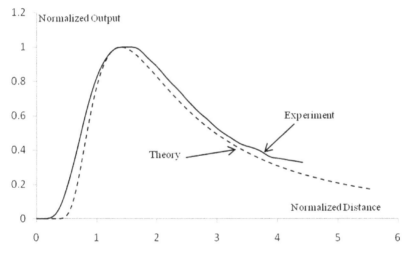

Fig. 16. Comparison between theoretical and experimental curves of the FODS with air medium in between the gap

the feasibility of our theoretical model. It is also observed that up to 0.3 (0.5) of separation for experimental (theoretical), light in transmitting fiber would be reflected back into itself and little or no light would be transferred to receiving fiber. This is then referred to as the blind region. As the distance increases, the reflected cone overlaps the receiving fiber core and hence the output intensity increases. This relation is continued until the entire face of receiving fiber is illuminated with the reflected light. This point is called optical peak and corresponds to maximum voltage. As the gap increases beyond this transition region, the intensity drops off following roughly an inverse-square law. The small discrepancy between the theoretical and experimental results is due to the noise sources such as shot noise and thermal noise, which are added to the value of the experimental results and are not calculated in theoretical analysis.

The experiments are also carried out to study the effect of k_1 and k_2 values as well as angle θ on the performance of FODS. Fig. 17 shows the normalized output power against displacement for the FODS at various k_1, k_2 and angle. Figs. 17 (a), (b) and (c) show the curves at 10°, 20°and 30° respectively with an air gap in between the displacement. By comparing the curves in Fig. 17, we understand that the performance of FODS is strongly depended on the fiber core size. The output power collected by receiving fiber is highest when the k_1 and k_2 values are set at 0.667 and 1, respectively. The inclination angle θ of two asymmetrical fiber core is also effected the sensor performance with the bigger inclined angle has a higher output sensitivity with a lower linearity range. Compared to the FODS with zero inclination angle, the sensitivity of the proposed sensor increased by 3.6, 8.5 and 16 times with the inclination angles of 10°, 20° and 30°, respectively. However, the corresponding linear range reduced by 67%, 55% and 33%, respectively. The performances of the proposed FODS are summarized as shown in Table 3.8. By using the k_1 and k_2 values of (0.667, 1), the maximum sensitivities of 0.2752 mV/mm, 0.3759 mV/mm and 0.7286 mV/mm are obtained at inclination angles of 10°, 20° and 30°, respectively. This sensitivity is higher compared to the previous work by Buchade and Shaligram (2006). The maximum linear ranges of 10.4mm, 7mm and 3mm are obtained at inclination angles of 10°, 20° and 30°, respectively for the FODS with k_1 and k_2 values of (0.667, 1).

Methods	Front slopes					
	Sensitivity (mV/mm)			Linearity Range (mm)		
(k_1, k_2)	10°	20°	30°	10°	20°	30°
(0.5, 0.5)	0.1345	0.1838	0.4761	1.5-3.5	0.4-2.1	0-0.7
(0.667,1)	0.2752	0.3759	0.7286	1 – 5.2	0.1-2.8	0 - 1
(1, 1)	0.1671	0.2224	0.5528	1.2-4.8	0.2-2.8	0 - 1
(1.5, 1)	0.1885	0.2745	0.6371	1.2-4.8	0.1-2.9	0 - 1
(2, 2)	0.0645	0.1201	0.1904	1.4-3.5	0.2- 2	0-0.8
	Back slopes					
	Sensitivity (mV/mm)			Linearity Range (mm)		
(k_1, k_2)	10°	20°	30°	10°	20°	30°
(0.5, 0.5)	0.0223	0.0479	0.1447	4.3-11.5	2.9-8.9	1.5-3.7
(0.667,1)	0.0675	0.1336	0.3224	6.6-17	3.2-10.2	1.3-4.3
(1, 1)	0.0472	0.0823	0.2296	5.6-14.8	3 - 9	1.2-4.3
(1.5, 1)	0.056	0.1155	0.2929	6.8-16	3.2-9.2	1.2-4.3
(2, 2)	0.0128	0.0389	0.0885	4.5-11.5	2.4-7.5	1 - 3

Table 6. The performances comparison of FODS with two asymmetrical inclined fibers

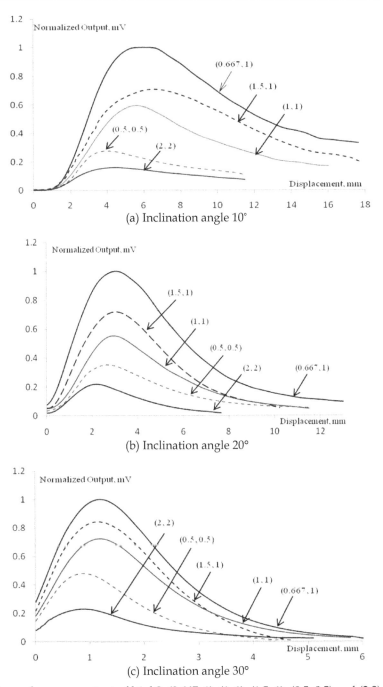

(a) Inclination angle 10°

(b) Inclination angle 20°

(c) Inclination angle 30°

Fig. 17. The performance output of k1, k2: (0.667, 1), (1, 1), (1.5, 1), (0.5, 0.5) and (2,2) in inclined angle (a) 10° (b) 20° (c) 30°

4. Applications of the FODS

Beside displacement measurement, the FODS can also be used to sense many other parameters such as temperature, pressure, refractive index, strain, mass and etc (Bechwith, 2000; Chang et al., 2008; Arellano-Sotelo, 2008). In this section, two different applications of FODS will be described.

4.1 Liquid refractive index detection using a FODS

In this study, the sensor probe is containing two pieces of fiber optics, one set connected to a light source and termed the transmitting fiber, and the other set connected to a photo detector (photodiode) and known as the receiving fiber. These two groups of fibers are bundled into a common probe to be used in a FODS. The FODS has a capability to measure physical quantities such as the displacement, vibration, strain, pressure, etc (Yasin et al., 2008; Yasin et al., 2009; Rastogi, 1997). However, the use of FODS sensors for detection of environmental refractive index change has not been fully explored. Refractive index sensing is important for biological and chemical applications since a number of substances can be detected through measurements of the refractive index. In the development of a liquid refractive index sensor (LRIS) (Suhadolnik et al., 1995; Chaudari and Shaligram, 2002; Yang et al., 2009; Nath et al., 2008; Kleizal & Verkelis, 2007), an intensity modulation in conjunction with multimode plastic fiber is the most suitable technique because of its non-contact sensing and many advantage properties are inherited by the multimode plastic fiber such as efficient signal coupling and being able to receive the maximum reflected light from target.

A FODS based refractive index measurement using a bundle fiber is first introduced by Suhadolnik et al. in 1995. Later on Chaudhari & Shaligram reported on study of LRIS at various types of optical sources. In our earlier work, a FODS was proposed based on two asymmetrical fibers for liquid refractive index measurement (Yang et al., 2009). In this section, a new LRIS is studied and demonstrated by using pair type of fibers bundled at various inclination angles.

The structure of proposed LRIS is shown in Fig. 18, which consists of a pair of transmitting and receiving fiber. We assume that the transmitting and receiving fibers have inclination angles of θ_1 and θ_2, respectively against the y-axis. The image of transmitting fiber is located opposite of the mirror with same distance. The central of the receiving fiber and the image of transmitting fiber are pointed as O' and O, respectively in Fig. 18. From the geometrical analysis of Fig. 18, the angle $\alpha = \sin^{-1}(NA/n)$ and $z_a = \dfrac{r_1}{\tan \alpha}$. Therefore, the following distances are given by;

$$AB = \sqrt{z_a^2 + r_{d1}^2} \sin\left(\tan^{-1}(\frac{r_{d1}}{z_a}) - \frac{\pi}{2} + \theta_1 \right)$$

$$O'C = 4r_{d1} \sin\theta_1 + 2x - AB - r_{d2}\sin\theta_2$$

$$OA = \sqrt{z_a^2 + r_{d1}^2} \cos\left(\tan^{-1}(\frac{r_{d1}}{z_a}) - \frac{\pi}{2} + \theta_1 \right)$$

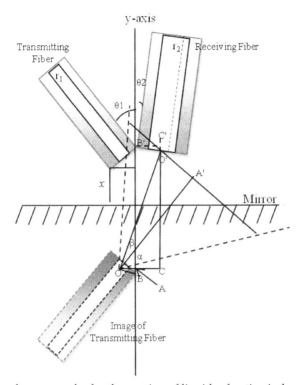

Fig. 18. Structure of sensor probe for the sensing of liquid refractive index.

$$OC = OA + r_{d2}\cos\theta_2$$

$$OO' = \sqrt{O'C^2 + OC^2}$$

where the NA is numerical aperture of transmitting fiber, n is refractive index of liquid, r_1 and r_2 are the core radius of transmitting fiber and receiving fiber while the r_{d1} and r_{d2} are the radius of transmitting fiber and receiving fiber, respectively and the x is the displacement between the sensor probe tip and reflector mirror.

Also from the geometrical analysis, the acceptance angle β of the light cone from the virtual point source O, is given by

$$\beta(z) = \tan^{-1}(\frac{O'C}{OC}) - \frac{\pi}{2} + \theta_1 \tag{29}$$

The intensity of the light emitted from the transmitting fiber can be well described with Gaussian distribution (Chang, et al., 2008) and is given by;

$$I(r, z) = \frac{2P_E}{\pi\omega^2(z)}\exp\left(-\frac{2r^2}{\omega^2(z)}\right) \tag{30}$$

where r is the radial coordinate, z is the longitudinal coordinate. $\omega(z)$ is the beam radius

and expressed as a function of z, $\omega(z) = \omega_0 \sqrt{1 + \left(\dfrac{z}{z_R}\right)^2}$. The waist radius ω_0 and Rayleigh

range z_R are the important parameters in the Gaussian Beam function and the detailed description can be found (Chang et al., 2008). Eq. (30) shows that the light intensity decays exponentially as it goes radially away from the center of the light circle. The radial coordinate r of Eq. (30) can be determined by;

$$r = OO' \sin \beta \qquad (31)$$

The longitudinal coordinate is the distance between the sensor probe tip and the virtual laser source point O and it can be determined

$$z = OO' \cos \beta \qquad (32)$$

For points situated in the far-field $z \gg z_R$, the beam radius of the virtual point source can be derived as

$$\omega(z) \approx z\alpha \qquad (33)$$

By the approximation,

$$r_1 = z_a \tan \alpha \approx z_a \alpha \qquad (34)$$

Based on the properties above, the power harnessed by the receiving fiber, P can be evaluated by integrating the Gaussian distribution function of Eq. (30) over the area of the of receiving fiber end surface, S_r

$$P(r,z) = \int_{S_r} I(r,z) dS_r \qquad (35)$$

where the core area of the receiving fiber is

$$S_r = \pi r_1^2 = \pi z_a^2 \alpha^2 \qquad (36)$$

By combining and substituting Eqs. (31), (32), (33) and (36) into the Eq. (35), finally the proposed LRIS response can be summarized as;

$$P_{(z,r)} = \frac{2P_E}{\pi \omega^2(z)} \exp\left(-\frac{2r^2}{\omega^2(z)}\right) \times S_r = \frac{2z_a^2 P_E}{z^2} \exp\left(\frac{2r^2}{z^2 \alpha^2}\right) \qquad (37)$$

This equation shows that the liquid refractive index response of sensor is a function of displacement x and refractive index n of surrounding medium while sensor probe is design of inclination angles of θ_1 and θ_2. Therefore, based on Eq. (37), the proposed LRIS can be used to detect the liquid refractive index where the sensor probe is immersed by the liquid to be measured.

The mathematical model of proposed LRIS is simulated by MATLAB programming. In the simulation, the wavelength of the laser source λ and numerical aperture NA is set at 594nm and 0.32, respectively. The fiber core radius r_1 and r_2 are set at 0.25mm and 0.5mm while the fiber diameters r_{d1} and r_{d2} are set at 0.5mm and 0.75mm, respectively. Fig. 19 shows the experimental set-up, which consists of a 594nm yellow He-Ne laser as a light source and a bundled fiber as a probe. The emitted laser light has an average output power of 3.0mW, beam diameter of 0.75mm and beam divergence of 0.92mRads. The external chopper modulates the light at a frequency of 200 Hz before it is launched into the transmitting fiber. The transmitting fiber transfers the modulated light from laser source to reflector mirror while the receiving fiber collects the reflection light to the detector. The sensor probe is mounted on the stage controlled by NewFocus Picomotor for the displacement measurement. Silicon detector measures the received light and converts it into electrical signal which is then denoised using the lock-in amplifier. During the measurement, the room temperature is keeping in 28°C to avoid the change of liquid refractive index.

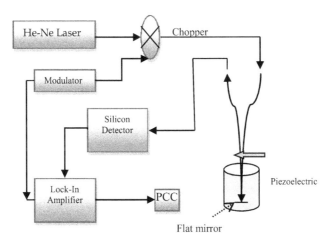

Fig. 19. Experimental set-up of the proposed liquid refractive index sensor

The simulation results are illustrated in Fig. 20. This figure shows the relationship between the sensor responses and probe inclination angles θ_1 and θ_2 in the measurement of three different refractive indices 1, 1.3 and 1.6 based on the probe inclined with the same angles of 0, 10° and 20°. In Fig. 5.8, the outputs powers are normalized in 1 and the displacements are simulated in *mm*. Three different group curves are showing three various sensors response based on the three various inclination angles. As seen in Fig. 20, it was found that the inclination angles θ_1 and θ_2 are reasonably affects the displacement curves profile and output power. The highest output power is almost 10 times of the lowest output power. The vertical dash lines are located in the displacements of 1.1 mm, 2.0 mm and 3.4 mm corresponding to the sensor probe inclination angles 20°, 10° and 0, respectively. In those positions, the sensor responses have the biggest output differentiation in the increase of refractive indices from 1.0 to 1.5. At those positions, the sensor output intensity increases almost linearly with the increase of the refractive index of the medium. From Fig. 20, we can observe that the increase the inclination angles improves the performance of the LRIS. The larger the inclination angles of θ_1 and θ_2, the better the performance of liquid refractive index response.

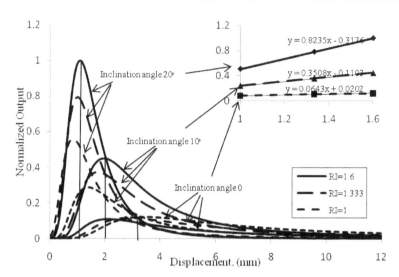

Fig. 20. Simulation results for the displacement at various inclination angles and refractive indices. Inset shows the maximum normalized output against the refractive index for inclination angles of 0°, 10° and 20°.

Inset of Fig. 20 shows the maximum normalized output of the sensor as a function of refractive index for various inclination angles. The normalized outputs were taken at the sensor probe positions of 1.1 mm, 2.0 mm and 3.4 mm for inclination angles of 20°, 10° and 0°, respectively, which is indicated by vertical dash lines in Fig. 20. It was found that the sensitivities of the sensor increases with the increment of probe inclination angle. As shown in the inset of Fig. 20, the highest sensitivity of 0.8235 is achieved by the use of probe with inclination angle of 20° which is almost 13 times higher than that in zero inclination. Fig. 21 shows the simulation curves of the LRIS at various inclination angles for the receiving and transmitting fibers when the refractive index of liquid is set at 1.3. It is clearly seen that the inclination angle of receiving fiber θ_1 has the stronger affect in the sensor output compared with the angle θ_2. As shown in Fig. 21, the highest output power is achieved by the inclination angles; $\theta_1 = 20°$ and $\theta_2 = 10°$. The lowest output power is observed when the inclination angles of θ_1 and θ_2 are set at 0° and 10°, respectively. These results show that the sensor sensitivity can be increased by increasing the inclination angle especially for θ_1. However, increasing the inclination is very difficult to be implemented in the experiment unless we can control the position of both fibers very precisely.

In our experiment, three different liquids: isopropyl alcohol, water and methanol are used as the surrounding medium at two conditions; zero inclination for both fibers and the same inclination angles of 10° for both fibers. The refractive index values for isopropyl alcohol, water and methanol are 1.377, 1.333 and 1.329, respectively. The sensor performance with air surrounding medium is also carried out for comparison purpose. During the experiment operation, the sensor probe is mounted onto the stage and the tank is fixed in the experiment table. The liquid in the tank is changed without moving the tank to ensure the accuracy of the measurement. The room temperature was kept at 28° to ensure that the refractive index of the liquid is maintained and only displacement parameter is changed in the experiment.

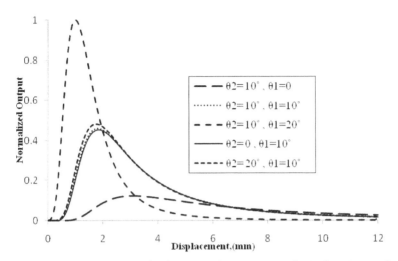

Fig. 21. The output power against displacement for various probe inclination angle combinations when the refractive index is set at 1.3.

Fig. 22 shows the displacement curve at various surrounding media when the inclination angles are set at 0° for both transmitting and receiving fibers. As shown in this figure, the normalized peak output power increases from 0.83 to 1.00 as the refractive index increases from 1.329 (methanol) to 1.377 (alcohol). It is also found that the displacement position for the peak output increases from 4.0 to 5.1 mm as the refractive index increases from 1.329 to 1.377. This is attributed the acceptance cone angle that increases with the refractive index increase. The larger acceptance angle provides a mean to collect more signal power. Fig. 23 shows the displacement curves when the inclination angles for both fibers are increased to 10°. As seen in Fig. 23, the peak power and its position increase with refractive index. The peak location of the curve also increases from 3.0 mm to 3.4 mm as the refractive index changes from 1.329 to 1.377. From these experimental results, it was found that the sensitivity of the sensor with 10° inclination of probe arrangement shows a higher sensitivity compared to that of the use of straight probe. The sensitivities are obtained at 0.11 mm^{-1} and 0.04 mm^{-1} for the sensors with 10° and 0° inclination angles respectively. This finding may be quite useful for chemical, pharmaceutical, biomedical and process control sensing applications.

4.2 The monitoring of liquid level using a FODS

Liquid level measurement is vital in many industry areas, finding the applications in such as fuel storage system, chemical processing, etc. Fiber based liquid level sensors are received in prefer of studies since they inherit many merits such as they are non-conductive, anti-erosion, and immune to electromagnetic interference. In past few years, several technical are employed to develop the fiber optic Liquid Level Sensors (LLSs). Fiber Bragg grating technology was carried out to sense the changing of liquid level by (Yun et al., 2007; Sohn & Shim, 2009; Dong & Zhao, 2010). Bending the multimode fiber for multi-points liquid level monitoring can be found in (Lomer et al., 2007). In these cases, the sensing elements are submerged inside the liquid to indicate the presence of liquid. These immersed sensing

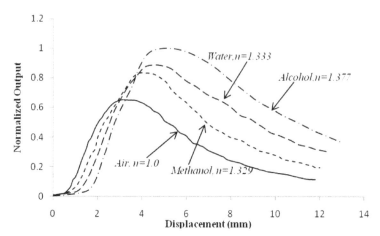

Fig. 22. Experiment results of the displacement curve at various liquid materials when the probe inclination angles are set at 0°.

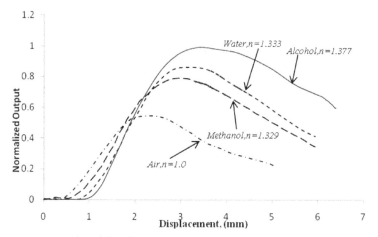

Fig. 23. Experiment results of the displacement curve at various liquid materials when the probe inclination angles for both transmitting and receiving fibers are set at 10°

elements have some limitations in their applications, such as in harsh environments, acid solutions, etc. Thereby, the fiber as sensing element for non-contact the measurement liquid seems good way to explore. There is one proposed sensor (Sohn and Shim, 2009) which employed one fiber bragg grating embedded inside the cantilever beam to connect with a float for indicating the level changing. This case is limited in its cost, surrounded temperature, and sensitivities.

In this section, a simply and cost efficiency intensity based LLS is proposed. A straightforward displacement sensor is employed; its sensing probe is engaged with a float which is contact with the measurement liquid to conduct the information of liquid. This simply setup transfers the sensing of liquid level into the measurement of displacement. It breaks down the limitations of sensitivities and measurement range which can be conquest by the selection of displacement sensors. Thus, in this study two type of displacement sensor are used to supply a flexible and compatible sensing of liquid level.

The basic sensing principle is the sensor probe displacement moving that causes from the moving of float during the liquid level change. A schematic setup of proposed fiber optic LLS is illustrated in Fig.24. It composed a He-Ne laser source light wavelength in 594nm (average output power 3.0mW, beam diameter 0.75mm and beam divergence 0.92mRads) which coupling the light into the transmitting fiber (2m length); a external chopper modulates the light at a frequency of 200Hz before launched into the transmitting fiber; a concave mirror (or flat mirror) is located at end of transmitting fiber and reflects the guided light of transmitting fiber into a receiving fiber which is bundled with the transmitting fiber. The receiving fiber guides the light into the photo detector which converts the light power into voltage. Meantime, the bundled fiber as sensor probe is fastened by an L cantilever beam which connects a float on the other side. The float is contact with the measurement liquid to indicate the information of liquid levels. A model SR-510 lock-in amplifier is connected with the chopper and photo detector. It plays the function of experiment data-acquisition system, matches the phase between the modulation light and modulator chopper and removes the noise generated by laser source, photo detector and amplifier.

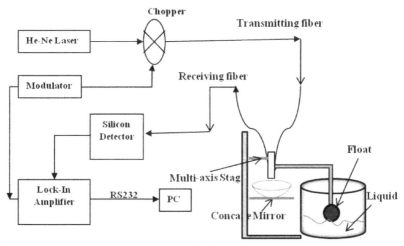

Fig. 24. Experiment setup of monitoring the liquid level

In the using of concave mirror as the reflector, the transfer function of proposed LLS is governed by parameters, namely focal length f and the diameter of the circular concave mirror D. The characteristic of the proposed sensor can be found the detailed theoretical analysis in paper (Yang et al., 2010). In consideration of the limited reflecting surface area and FL of the concave mirror, the transfer function of the proposed sensor can be given as

$$P(u) = \frac{2z_a^2 P_E \left\{ 1 - \exp\left(-\frac{D^2}{2(u+z_a)^2 \theta^2} \right) \right\}}{\left[u - \frac{(u+z_a)f}{u-f+z_a} \right]^2 \left(\frac{u+z_a}{f} - 1 \right)^2} \exp\left(-\frac{8z_a^2}{\left[u - \frac{(u+z_a)f}{u-f+z_a} \right]^2 \left(\frac{u+z_a}{f} - 1 \right)^2} \right) \qquad (38)$$

where P_E is the emitting power of laser source, u is the distance between the sensor probe tip and concave mirror, $\theta = \sin^{-1}(NA)$ and NA is the numerical aperture of the transmitting fiber. By using the flat mirror as the reflector, the transfer function of proposed sensor was analyzed in (Lim et al., 2009) detailed. It is given by

$$P(z) = \frac{2(k+1)^2}{\xi^2}\exp\left(1 - \frac{2(1+k)^2}{\xi^2}\right) \tag{39}$$

The core radial ratio k is the ratio of the transmitting fiber core radius to the receiving fiber core radius, ξ is the normalized displacement.

From the transfer functions of Eq. (38) and (39), it can be seen that the sensor response is only influenced by the focal length f and the diameter of the circular concave mirror D in the concave mirror as reflector and by the core radial ratio k for that of flat mirror. As such, the proposed LLS has the widely compatible for the variants type of FODSs, which can escapes the trade-off design in the selection of sensitivity and measurement range. Furthermore, according to the Eqs. (38) and (39), there are no more parameters can affect the sensor response, hence, the proposed LLS is independency for surrounded environments.

Fig. 25 depicts the experiment result when the liquid level is moving upward. In this experiment, the proposed LLS is composed two plastic fiber bundled parallel as sensor probe, a concave mirror with focal length 10mm and diameter 24mm as reflector. From Fig. 25, it can be seen that the sensor output intensity is modulated by the moving of liquid level in six variants slops. According to this, hence, the proposed sensor can achieve the continue monitoring of liquid level in the multi-points. Totally six monitoring points are shown in Fig. 25, each point is represented each slop which can be used as the each level of liquid measurement. These six monitoring points are located in the total measurement range of 25mm, which can provide variants sensitivities and measurement range. The performances of these monitoring points are summarized in table 7.

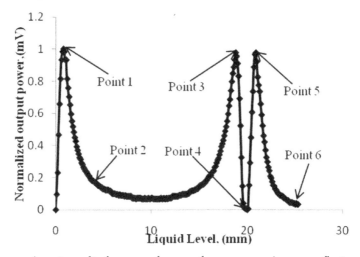

Fig. 25. The experiment result of proposed sensor for concave mirror as reflector and two plastic fibers bundled parallel

The second experimental is carried out by the sensor probe consisted with two bundled parallel and a flat mirror as reflector. On the other hand, another experimental is repeated by two asymmetrical plastic fiber bundled parallel as sensor probe and a flat mirror as reflector. Both of responses are illustrated in Fig. 26, where each of the curves are displayed two various slops in the measurement range of 3mm and 3.5mm, respectively. From the observation of Fig. 26, two various monitoring points are given from both curves as shown. They are occupied the measurement ranges of 3mm and 3.5mm, respectively. In Fig. 26, the higher sensitivity is obtained from the curve of k=1 and the bigger measurement range is supplied by the curve of k=0.5. The output curves in Fig. 26 are shifted to the right along the further of liquid level as the value of k decreases. Therefore, the sensitivities of front slope and back slope of the curves are decreasing with decreasing the values of k. Mean times, the linearity ranges of curves are enlarged with decreasing the values of k. The sensitivities and linearity ranges of Fig. 26 are summarized in Table 7.

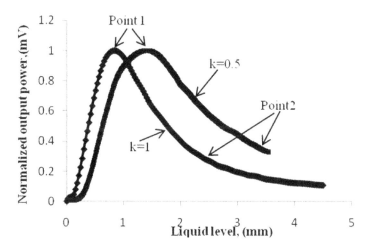

Fig. 26. The experiment results of various ratios of fibers bundled as sensor probe.

Table 7 tabulates the performances of LLS for three different experiments. From the results summarized in table 7, it can be seen that the highest sensitivity is 1.4533mV/mm when its measurement range is 0.84mm. The largest measurement range 4.8mm is obtained from the monitoring point 3 for the concave mirror as reflector. In the experiment of concave mirror as reflector, the monitoring points 1, 4, and 5 have the higher sensitivities and shorter measurement ranges than that of 2, 3, and 6. In the experiment of flat mirror as reflector, the sensitivities for k=0.5 are decreased by 52.5% and 14.8% for of front slope and back slope, respectively, if compared to k=1. The measurement ranges of front slope and back slope are increased by 46.2% and 18.8% if compared with the k=1. From the analysis, by using a particular combination of fiber core diameters and reflector, two different k values, ten different sensitivity and linear range selections can be achieved by using the proposed method. Thus, the proposed fiber optic LLS has wide compatibility for other type of displacement sensor to avoid the trade-off selection of sensitivity and measurement range which is always encountered in the design of conventional fiber optic LLS.

Methods	Monitoring point	Sensitivity (mV/mm)	Measurement range (mm)
Concave mirror k=1	Point 1	1.4026	0-0.8
	Point 2	0.2649	0.8-3.8
	Point 3	0.1625	14-18.8
	Point 4	1.030	18.9-20
	Point 5	1.3015	20.1-20.9
	Point 6	0.1821	21-25
Flat mirror k=0.5	Point 1	0.9529	0.1-1.4
	Point 2	0.3443	1.4-3.5
Flat mirror k=1	Point 1	1.4533	0-0.84
	Point 2	0.3951	0.84-3

Table 7. The Sensitivities and measurement ranges of proposed sensor

In the literature of (Sohn and Shim, 2009), the wavelength-shift sensitivity is around 0.15nm/cm and the resolution of intensity modulated the liquid level is approximated 0.1dB/cm are achieved. However, these sensitivities cannot satisfy the high precision measurements and the LLS proposed in (Sohn and Shim, 2009) is high costly. In this paper, the highest sensitivity is achieved by using a probe with two fibers bundled and flat mirror as reflector, it is around 100 times higher than that of (Sohn and Shim, 2009) and is 20 times lower in the cost.

5. Conclusion

In this chapter, the performance of various FODSs is investigated theoretically and experimentally. A beam-through FODS has been demonstrated for sensing of lateral and axial displacements, which shows that the usage of smaller core normally shows a better sensitivity and resolution with the expense of the smaller linear range. Various reflective FODSs are then proposed to provide a possibility for the development of the cheap, simple, sensitive and wide dynamic range sensor systems. Compared with the conventional sensor with only one receiving fiber, FODS with two receiving fibers is observed to have a better linearity range. The performance of FODS with asymmetrical bundled fiber is also theoretically and experimentally demonstrated in this chapter. The effect of different core radial ratios (CRRs) on the performance of the sensor is investigated in terms of dynamic range and sensitivity. The experimental results are almost in agreement with the theoretical results. The location of the maximum output is shifted toward the right along the axis of displacement as the value of k increases. Besides, the linear range for both front and back slopes increases with the value of k. To improve the sensitivity, a FODS with two inclined asymmetrical fiber has been proposed. This FODS has been applied in liquid refractive index sensor. An extra-low cost, ultra-high sensitivity, and wide compatibility liquid level fiber optic sensor has been demonstrated as another application for the FODS. A float is touched with the measurement liquid to translate the information of liquid change to the sensor probe displacement moving where an L cantilever beam is fixed in between, the moving of sensor probe finally results the output intensity is modulated. All the theoretical and experimental results are capable of offering quantitative guidance for the design and implementation of a practical FODS.

6. References

Arellano-Sotelo, H.; Barmenkov, Yu. O.; Kir'yanov, A.V. (2008). The use of erbium fiber laser relaxation frequency for sensing refractive index and solute concentration of aqueous solutions. *Laser Physics Letters*, Vol. 5, No. 11, pp. 825-829.

Bechwith, M.L. (2000). Mechanical measurement. Addison Wesley Longman.

Bergamin; Cavagnero, G.; Mana, G. (1993). A displacement and angle interferometer with subatomic resolution. *Rev Sci Instrum*, 64, 3076-3081.

Bock, W.J.; Nawrocka, M.S.; Urbanczyk, W. (2001). Highly sensitive fiber optic sensors for dynamic pressure measurement. *IEEE Trans. Instrum. Meas.*, Vol. 50, No. 5, pp. 1085-1088.

Buchade, P.B. & Shaligram, A.D. (2006). Simulation and experimental studies of inclined two fiber displacement sensor. *Sensor and Actuators A*, Vol. 128, pp. 312- 316.

Buchade, P.B. & Shaligram, A.D. (2007). Influence of fiber geometry on the performance of two-fiber displacement sensor. *Sensors and Actuators A*, 136, 199- 204.

Cao, M.H.; Chen, Y.; Zhou, Z.; Zhang, G. (2007). Theoretical and experimental study on the optical fiber bundle displacement sensors. *Sensor and Actuators A*, Vol. 136, pp. 580-587.

Chang, J.; Wang, Q.; Zhang, X.; Ma, L.; Liu, T.; Wang, Q.; Liu, Z.; Zhang S.; Ding, S. (2008). Single-end vibration sensor based on an over-coupled fiber-loop reflector. *Laser Physics*, Vol. 18, pp. 452-454.

Chaudhari, A.L. & Shaligram, A.D. (2002). Multi-wavelength optical fiber liquid refracto-metry based on intensity modulation", *Sensor and Actuators A*, Vol.100, pp.160- 164.

Cui, C.; Wang, Q.; Hu, X.; Yu, Y.; Zhao, S. (2008). Simulation of spin-axis position measurement of superconducting sphere rotor by fiber optic sensor. *IEEE Trans Appl Supercond.*, Vol 18, pp. 836-839.

Dong, X.W. & Zhao, R.F. (2010). Detection of liquid level variation using a side polished fiber bragg grating. *Optics & laser technology*, 42, 214-218.

Elasar, J.; Selmic, S.; Tomic, M.; Prokin, M. (2002). A fiber optic displacement sensor for a cyclotron environment based on a modified triangulation method. *J. Opt. A: Pure Appl. Opt.*, Vol. 4, pp. 347–355.

Faria, J.B. (1998). A theoretical analysis of the bifurcated fiber bundle displacement sensor. *IEEE Transactions on instrumentation and measurement*, Vol. 47, pp. 742-747.

Kleiza1, V. & Verkelis, J. (2007). Some Advanced Fiber-Optic Amplitude Modulated Reflection Displacement and Refractive Index Sensors. *Nonlinear Analysis: Modelling and Control*, Vol. 12, No. 2, pp. 213–225.

Ko, W.H.; Chang, K.M.; Hwang, G.J. (1995). A fiber optic reflective displacement micrometer. *Sensor Actuators A*, Vol. 49, pp. 51–55.

Kulchin, Y.N.; Vitrik O.B.; Dyshlyuk A.V.; Shalagin, A.M.; Babin, S.A.; Vlasov, A.A. (2007). Applications of optical time domain reflectometry for the interrogation of fiber bragg sensors. Laser Physics, Vol. 17, pp. 1335-1339.

Laurence Bergougnous, Jacqueline Misguich-Ripault, and Jean-Luc Firpo, (1998). Characterization of an optical fiber bundle sensor. *Rev. Sci. Instrum.* 69, 1985–1990.

Lim, K.S.; Harun, S.W.; Yang, H.Z.; Dimyati, K.; Ahmad, H. (2009). Analytical and experimental studies on asymmetric bundle fiber displacement sensors. *J. of Modern Optics*, Vol. 56, pp.1838-1842.

Lomer, M.; Arrue, J.; Jauregui, C.; Aiestaran, P.; Zubia, J.; Lopez-Higuera, J.M. (2007). Lateral polishing of bends in plastic optical fibers applied to a multipoint liquid level measurement sensor. *Sensors and Actuators A*, 137, 68-73.

Miclos, S. & Zisu, T. (2001). Chalcogenide fibre displacement sensor. J. Optoelectron Adv. Matter., Vol. 3, pp. 373-376.

Murphy, A.M.; Gunther, M.F.; Vengsarkar, A.M.; Claus, O.R. (1991). Quadrature phase-shifted extrinsic Fabry-Perot optical fiber sensor. *Opt. Lett* ., 16, 273.

Murphy, M.M. & Jones, G.R. (1994). A variable range extrinsic optical fiber displacement sensor. *Pure Appl. Opt.*, Vol. 3, pp. 361–369.

Nalwa, S. (2004). Polymer optical fibers. California: American Scientific Publishers.

Nath, P.; Singh, H.K.; Datta, P.; Sarma, K.C. (2008). All-fiber optic sensor for measurement of liquid refractive index. *Sensor and Actuators A*, Vol.148, pp.16- 18.

Oiwa T. & Nishitani, H. (2004). Three dimensional touch probe using three fiber optic displacement sensor. *Meas. Sci. Technol.*, Vol. 15, pp. 84–90.

Rastogi, P.K. (1997). *Optical Measurement Techniques and Applications*. Artech House, Inc. Boston, London.

Saunders, C. & Scully. (2007). Sensing applications for POF and hybrid fibres using a photon counting OTDR. Meas. Sci. Technol., Vol. 18, pp. 615-622.

Shimamoto, A. & Tanaka, K. (2001). Geometrical analysis of an optical fiber bundle displacement sensor. Applied Optics, Vol. 35, Issue 34, pp. 6767-6774.

Sohn, K.R. & Shim, J.H. (2009). Liquid level monitoring sensor system using fiber bragg grating embedded in cantilever. *Sensors and Actuators, A*, 152, 248-251.

Spooncer, R.C.; Butler, C.; Jones, B.E. (1992). Optical fibre displacement sensors for process and manufacturing applications. Opt. Eng. 31, 1632-1637.

Suhadolnik; Babnik, A.; Mo2ina,J. (1995). Optical fiber reflection refractometer. *Sensor and Actuators B*, Vol. 29, pp.428- 432.

Van Etten W. & Van der Plaats J. (1991). Fundamentals of optical fiber communi-cations. Prentice-Hall, London.

Wang, D.X.; Karim, M.A.; Li, Y. (1997). Self referenced fiber optic sensor performance for micro displacement measurement. Opt. Eng., Vol. 36, pp. 838–842.

Yang, H.Z.; Lim, K.S.; Harun, S.W.; Dimyati, K.; Ahmad, H. (2010). Enhanced Bundle Fiber Displacement Sensor Based on Concave mirror. *Sensors and Actuators A*, 162, 8-12.

Yang, H.Z.; Harun, S.W.; Ahmad, H. (2009). Fiber Optic Displacement and Liquid Refractive Index Sensors with Two Asymmetrical Inclined Fibers. *Sensor and Transducer*, Vol.108, pp. 80-88.

Yasin,M.; Harun, S.W.; Pujiyanto, Ghani, Z.A.; Ahmad H. (2010). PerformanceComparison between Plastic-Based Fiber Bundle and Multimode FusedCoupler as Probes in Displacement Sensors. *Laser Physics*, Vol. 20, No. 10, pp. 1890–1893.

Yasin, M.; Harun, S.W.; Abdul-Rashid, H.A.; Kusminarto; Karyono; Ahmad, H. (2008). The performance of a fiber optic displacement sensor for different types of probes and targets. *Laser Physics Letters*, Vol. 5, No. 1, pp. 55-58.

Yasin, M.; Harun, S.W.; Samian; Kusminarto; Ahmad, H. (2009). Simple design of optical fiber displacement sensor using a multimode fiber coupler. *Laser Physics*, Vol. 19, No.7, pp. 1446-1449.

Yun, B.F.; Chen, N.; Cui, Y.P. (2007). Highly sensitive liquid level sensor based on etched fiber bragg grating. *IEEE Photonic Technology Letters*, 19, 1747-1749.

4

High-Birefringent Fiber Loop Mirror Sensors: New Developments

Marta S. Ferreira, Ricardo M. Silva and Orlando Frazão
INESC Porto, Porto
Portugal

1. Introduction

The fiber loop mirror based on a ring configuration is one of the devices in optical fibers most used in communications and sensors (Mortimore, 1988). The device is formed when the two output ports of a directional coupler are spliced. In this configuration, the two waves at the coupler outputs travel in opposite directions but following the same optical paths in an optical fiber ring, which assures constructive interference as the waves re-enter the coupler. Thus, all light is reflected to the input port containing losses essentially in the fiber, the splice region and the coupler. Due to this "mirror" characteristic, the device is frequently used in the formation of resonant cavities in optical fiber lasers (Urquhart, 1989). A structure of this kind, with interesting properties, occurs when the ring contains a section of fiber with high-birefringence. When this happens, an interference pattern is generated. It depends only on the fiber birefringence and length, being independent of the remaining ring extension (Fang & Claus, 1995).

This chapter provides an overview of the state-of-the-art, birefringence concepts, and new developments of high-birefringence fiber loop mirror configurations that can be used as sensing elements.

2. State-of-the-art

In this section, a brief review about the state-of-the-art using high-birefringent fiber loop mirror (Hi-Bi FLM) configurations is described. In order to give a clear view about the evolution of Hi-Bi FLM setups, a sum up with the years of publications and the most important breakthroughs are shown in table 1.

The first Hi-Bi fiber Sagnac loop sensor for temperature measurement was proposed in 1997. The configuration, schematically shown in Figure 1 a), consisted of a 3 dB simple polarization maintaining coupler, constituted by a bow-tie fiber. By cross splicing the output ports at $\pi/2$, they guaranteed that both port lengths were different, forming an unbalanced Sagnac loop interferometer.

The temperature sensor could operate both in transmission and reflection, and its functioning did not depend on the light polarization at the input port. When compared with other temperature optical sensors (fiber Bragg gratings), this configuration presented higher sensitivity (Delarosa et al., 1997; Starodumov et al., 1997). In the same year, a strain sensor

Year	Breakthrough of Hi-Bi FLM configurations
1997	Temperature sensor (Starodumov et al., 1997); temperature sensing in NIR (Delarosa et al., 1997); strain sensor (Campbell et al., 1997)
1999	Strain sensor (Campbell et al., 1999)
2004	Temperature PCF sensor (Kim, D. H. & Kang, 2004); temperature insensitivity using PCF (Zhao et al., 2004)
2005	Interrogation system using PCF (Yang et al., 2005); displacement sensor (Liu et al., 2005)
2006	Liquid level sensor (Dong, B. et al., 2006); FBG interrogation system (Silva et al., 2011); LPG/Hi-Bi FLM (Frazao et al., 2006b)
2007	Review of Hi-Bi FLM sensors (Frazao et al., 2007a); strain PCF sensor (Dong, X. Y. et al., 2007; Frazao et al., 2007b) ; chemical etching (Frazao et al., 2007c); concatenated FLM (Frazao et al., 2007d)
2008	Temperature Erbium Hi-Bi fiber (Frazao et al., 2008a); refractive index sensor (Frazao et al., 2008b); multiparameter sensor using side-hole fiber (Frazao et al., 2008c); pressure PCF sensor (Fu et al., 2008); FBG/Hi-Bi FLM (Zhou et al., 2008); current sensor (Marques et al., 2008)
2009	Multiplexing Hi-Bi FLM (Fu et al., 2009); strain and temperature discrimination using two Hi-Bi fibers (Han et al., 2009); hollow-core PCF sensor (Kim, G. et al., 2009); holey fiber filled with metal indium (Kim, B. H. et al., 2009)
2010	Torsion PCF sensor (Kim, H. M. et al., 2010); curvature PCF sensor (Gong et al., 2010); pressure-induced SMF (Jin et al., 2010); long distance remote interrogation system (Lee et al., 2010); small core PCF (Andre et al., 2010); intensity strain sensor (Qian et al., 2010); suspended twin-core (Frazao et al., 2010); displacement PCF sensor (Zhang et al., 2010)
2011	Hi-Bi FLM with an output port probe (Frazao et al., 2011); LPG/Sagnac Loop (Kang et al., 2011); curvature PCF sensor (Hwang et al., 2011); Hi-Bi photonic bandgap Bragg fiber (Ferreira et al., 2011); transverse mechanical load (Zu et al., 2011a); PCF FLM filled with alcohol (Qian et al., 2011); PCF FLM with core offset (Dong, B. et al., 2011); magneto-optic modulator (Zu et al., 2011b); tapered FLM (Zibaii et al., 2011)

Table 1. Hi-Bi FLM configurations chronology.

based on a Hi-Bi fiber Sagnac ring configuration was investigated (see Figure 1 b) (Campbell et al., 1997). In 1999, the same authors reported similar work using a frequency-modulated continuous wave (Campbell et al., 1999).

In 2004, temperature dependence of a polarization maintaining (PM) photonic crystal fiber (PCF) in a Sagnac loop interferometer was analyzed. Comparing with a standard PM fiber, they obtained reduced temperature sensitivity (Kim, D. H. & Kang, 2004). A temperature insensitive Hi-Bi FLM using a photonic crystal fiber was demonstrated. For low-temperatures (25°C to 85°C), a wavelength spacing variation of 0.05 pm/°C was achieved (Zhao et al., 2004).

In 2005, the characteristics of a Hi-Bi FLM composed of a standard 3 dB fiber coupler with one or more sections of Hi-Bi fibers were discussed, as can be seen in Figure 1 c). Displacement and temperature sensors were theoretically and experimentally analyzed.

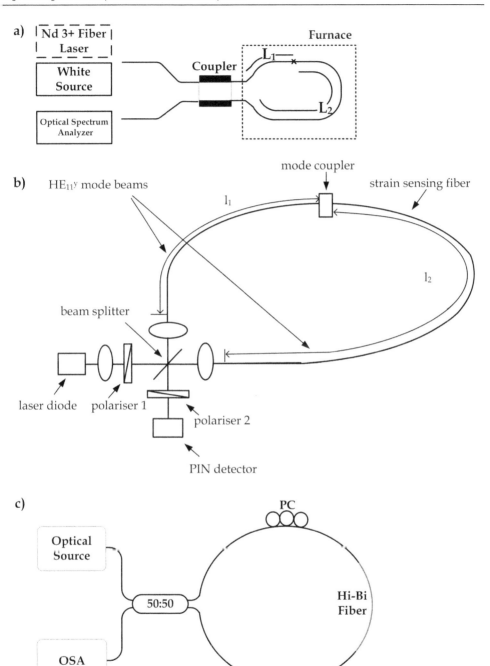

Fig. 1. a) First temperature sensor using a Sagnac interferometer, b) First strain sensor and c) conventional Hi-Bi FLM.

Using the characteristics of the Hi-Bi FLM, a strain-temperature sensor was proposed and demonstrated as an interrogation system for new applications (Liu et al., 2005). A temperature insensitive fiber Bragg grating (FBG) sensor interrogation system, which used a PCF in a Hi-Bi FLM was proposed and experimentally demonstrated (Yang et al., 2005).

In 2006, a liquid-level Hi-Bi FLM sensor based on a uniform-strength cantilever beam was presented (Dong, B. et al., 2006). The advantages of this sensor included its simple structure, high sensitivity, low-cost and repeatability. A FBG interrogation scheme based on a Hi-Bi FLM for strain-temperature discrimination was reported (Frazao et al., 2006a). The configuration improvement was the use of a Hi-Bi FLM to interrogate an array of FBGs. The same author proposed an alternative solution for simultaneous measurement of strain and temperature, using a long-period fiber grating (LPG) combined with a Hi-Bi FLM (Frazao et al., 2006b).

In 2007, a fiber-optic strain sensor was demonstrated by using a short length of PCF inserted in a Hi-Bi FLM (Dong, X. Y. et. al., 2007). This strain sensor presented temperature insensitivity. The investigation was done using an acrylate coated PCF and other uncoated. Strain and temperature measurements were made to verify their sensibilities (Frazao et al., 2007b). The uncoated PCF revealed insensitivity to temperature. Another work made by the same author (Frazao et al., 2007c) was the birefringence control of a Hi-Bi fiber (bow-tie) under chemical etching. This fiber was inserted in a FLM and the optical response was measured while chemical etching of the bow-tie fiber was taking place. A different study made, was a configuration for simultaneous strain and temperature discrimination using two concatenated Hi-Bi FLMs (Frazao et al., 2007d). By monitoring the wavelength and optical power, it was possible to obtain both measurements simultaneously.

In 2008, an optical current sensor based on Hi-Bi FLM was reported (Marques et al., 2008). An elliptical cladding fiber was coated with a metal and inserted as a sensing element in the FLM. A novel intrinsic fiber optic pressure sensor, based on a PM-PCF section inserted between the output ports of the Hi-Bi FLM was presented (Fu et al., 2008). This sensing head did not require polarimetric detection as the ones reported so far, and through wavelength shifts they obtained a sensitivity of 3.42 nm/MPa for a 58.4 cm long PM-PCF. A setup where the Hi-Bi section was combined with a FBG, enabling the simultaneous measurement of strain and temperature was presented (Zhou et al., 2008). The obtained sensing resolution was of ±1 °C for temperature and ±21 με for strain. A Hi-Bi PANDA Erbium-doped FLM was presented for the first time. In this case, a temperature coefficient sensitivity of -2.22 nm/°C was achieved (Frazao et al., 2008a). Another report by the same author was a Hi-Bi D-type FLM used as refractometer (Frazao et al., 2008b). A chemical etching was applied to the D-type fiber and presented high sensitivity to refractive index. In the same year, a configuration for simultaneous measurement of multiparameters by using a PM side-hole fiber section in the FLM was proposed (Frazao et al., 2008c). Different parameters were analyzed, like torsion, temperature and longitudinal strain.

In the year of 2009, high temperature sensitivity was achieved when a section of birefringent holey fiber filled with metal indium was introduced in a FLM (Kim, B. H. et al., 2009). The analysis was done for wavelength and birefringence responses to temperature, and sensitivities of -6.3 nm/K and -3.3×10⁻⁶ K⁻¹ were respectively obtained. On the other hand, an elliptical hollow-core photonic bandgap fiber was fabricated and tested in a FLM configuration (Kim, G. et al., 2009), measuring strain and temperature responses. Later on, simultaneous measurement of temperature and strain was performed by using a PM-PCF with Erbium-doped fiber incorporated (Han et al., 2009). Finally, three different multiplexing schemes for PCF FLMs were presented (Fu et al., 2009).

In 2010, a temperature-insensitive displacement sensor, based on Hi-Bi PCFLM was proposed (Zhang et al., 2010). This sensor presented a displacement sensitivity of 0.28286 nm/mm and good stability through a temperature range from 40 °C to 109.5 °C. A low-birefringence PCF FLM as a curvature sensor was reported for the first time (Gong et al., 2010). A sensitivity of -0.337 nm for a range between 0-9.92 m⁻¹ was achieved. This sensor was also sensitive to temperature, transverse load and twist, thus being a potential multifunction sensor. A temperature sensor using a section of pressure-induced birefringent SMF FLM was firstly presented (Jin et al., 2010). A sensitivity of 0.65 nm/°C was reported. A temperature-independent strain sensor based on intensity measurements was developed (Qian et al., 2010). A small core microstructured fiber segment in the FLM for the simultaneous measurement of strain and temperature was used (Andre et al., 2010). Besides Hi-Bi, this configuration also showed intermodal interference, resulting in a complex channeled spectrum. With a sensing head composed only by a fiber section, resolutions of ±1.5 °C and ±4.7 µε for temperature and strain were respectively obtained. A torsion sensor with temperature and strain independence, using a suspended twin-core fiber to form the Hi-Bi FLM was investigated (Frazao et al., 2010). The variations in amplitude caused by applying torsion on the sensing head were analyzed by the Fast Fourier Transform technique. Another torsion sensor was done by placing a PCF in a FLM and demonstrated its temperature insensitivity (Kim, H. M. et al., 2010). The variation of reflectivity for long-distance remote FGB cavity sensors using a Hi-Bi FLM was analyzed and demonstrated (Lee et al., 2010).

In 2011, a curvature sensor with a new Hi-Bi PCF design spliced inside the loop mirror was demonstrated (Hwang et al., 2011). The proposed PCF had two large air holes in the outer cladding region that gave rise to core ellipticity during the fabrication process. The curvature sensor is based on the PCF Sagnac interferometer and its sensitivity depended on the bending directions with respect to the large-air-hole alignment. A simultaneous measurement of strain and temperature based on cladding-mode-resonance in PM PCF loop mirror was reported (Dong, B. et al., 2011). By introducing a mode field diameter mismatch at a splicing joint of an inbuilt SMF-PMPCF-SMF structure connected in a Sagnac configuration, an intermodal interference was constructed by the excited cladding modes and the polarization modes of the PMPCF. An optical fiber transverse mechanical load sensor using a short section of solid core PCF which was inserted inside a FLM was reported (Zu et al., 2011a). Two new configurations of Hi-Bi FLMs with an output port probe were proposed (Frazao et al., 2011). Both configurations used two couplers spliced in between, with unbalanced arms and one output port was used as the probe sensor. The difference between them was the location of the Hi-Bi fiber section length: either between the two couplers (first new configuration) or spliced at the output port probe (second new configuration). These configurations were compared with the conventional Hi-Bi FLM when strain was applied and showed similar sensitivities. The theoretical model of these two configurations was studied by (Silva et al., 2011). A compact temperature sensor based on a

FLM combined with an alcohol-filled Hi-Bi PCF was investigated (Qian et al., 2011). A magneto-optic modulator with a magnetic fluid film inserted into an optical fiber Sagnac interferometer was proposed (Zu et al., 2011b) The magnetic fluid exhibited variable birefringence and Faraday effect under external magnetic field that led to a phase difference and polarization state rotation in the Sagnac interferometer. A Hi-Bi FLM with a LPG inscribed in PMF was proposed and experimentally demonstrated for simultaneous measurement of strain and temperature (Kang et al., 2011). This was due to the different responses of the LPG and the Sagnac interferometer observed. A Hi-Bi photonic bandgap Bragg FLM configuration for simultaneous measurement of strain and temperature was reported (Ferreira et al., 2011). The group birefringence and the sharp loss peaks were observable in the spectral response. Since the sensing head presented different sensitivities for strain and temperature measurands, these physical parameters were discriminated by using the matrix method. A single-mode non-adiabatic tapered optical fiber sensor inserted in a FLM enabling to tune its sensitivity towards refractive index was reported (Zibaii et al., 2011).

3. Birefringence concept

The Hi-Bi FLM consists in a coupler and a Hi-Bi fiber section as a sensing head. The pattern fringe is obtained through the fiber characteristics. The birefringence is guaranteed in the fabrication process through the geometrical or stress effects. The geometrical effect is easily obtained when the core is non-circular, for example, the elliptical core fiber. The stress effect can be obtained by applying mechanical tension around the core, as occurs in the case of PANDA or bow-tie fibers.

The birefringence effect in optical fibers was studied considering the phase and group modal birefringence. The phase modal birefringence is defined as:

$$B(\lambda) \equiv n_x - n_y = \frac{\lambda}{2\pi}\left(\beta_x(\lambda) - \beta_y(\lambda)\right),$$ (1)

where β_i ($i=x, y$) is the propagation constant associated to each fundamental mode of polarization. This birefringence is associated to the beat length, L_B, through the relation:

$$L_B = \frac{\lambda}{B(\lambda)}$$ (2)

The group modal birefringence, $G(\lambda)$, is given by:

$$G(\lambda) = \frac{d\beta_x(\lambda)}{dk} - \frac{d\beta_y(\lambda)}{dk} = B(\lambda) - \lambda\frac{dB(\lambda)}{d\lambda}$$ (3)

where $k = 2\pi / \lambda$. The differential phase between two polarization modes after propagating through an extension L of fiber is given by:

$$\phi = \frac{2\pi L B(\lambda)}{\lambda}$$ (4)

resulting in a channeled type interferometric spectral response, as illustrated in Figure 2.

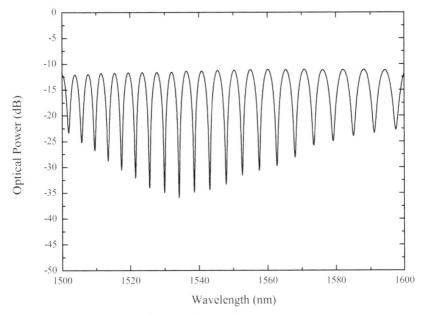

Fig. 2. Typical spectral response of an interferometric structure.

Deriving the phase ϕ with respect to λ, the following equation is obtained:

$$\frac{d\phi}{d\lambda} = \frac{2\pi L}{\lambda^2}\left(\lambda\frac{dB(\lambda)}{d\lambda} - B(\lambda)\right) = -\frac{2\pi L}{\lambda^2}G(\lambda) \qquad (5)$$

If $\Delta\lambda$ is the channeled spectrum period, corresponding to a phase variation $d\phi = 2\pi$, the expression obtained for the group modal birefringence is:

$$G(\lambda) = -\frac{\lambda^2}{\Delta\lambda L}. \qquad (6)$$

4. New developments

New developments have been proposed using the FLM in novel designs. Recently, three new Hi-Bi fiber loop mirror with output port probe configurations were demonstrated. These configurations present new solutions as sensing elements. One of them is to use the output port as optical refractometer without the Hi-Bi fiber section or conventional strain/temperature sensors when the probe is a Hi-Bi fiber section.

A torsion sensor with insensitivity to temperature and strain measurement is also demonstrated. In this case the conventional optical coupler is changed by a Hi-Bi coupler where the optical path difference is obtained when the two output arms are unbanlanced. These two new configurations are presented in the following sections.

4.1 Hi-Bi FLM with output probe

Three novel configurations of Hi-Bi FLMs are schematically shown in Figure 3. They consist of a loop of optical fiber formed between the output ports of two directional couplers along with an output port probe.

The first configuration proposed has a polarization controller and a Hi-Bi fiber section located between the optical couplers, (Figure 3 a)). The mode of operation is similar to the conventional FLM. In the first coupler, the light is divided in two waves that travel between the two couplers with asymmetric arms and re-couple in the output port of the second directional coupler. No interference is obtained in the output port of the second coupler because the arms between the two couplers are unbalanced. A silver mirror was fabricated in the end of the fiber tip to re-inject the light into the second coupler. Here the two waves are divided in four waves and re-injected into the first coupler.

Only the two waves that travel in different paths in the two couplers section in the download and upload propagation have a path imbalance associated with the Hi-Bi fiber and, therefore, it is obtained a channeled spectrum with a periodicity similar to what is expectable when considering the standard configuration. The other two waves accumulate a substantial path imbalance and therefore the corresponding channeled spectrum has a very small periodicity, not resolvable by the OSA used. Therefore, the fringe pattern observed in the OSA comes out from the phase difference between the polarization states of the Hi-Bi fiber length. In this case, this configuration was characterized as optical refractometer (Frazao et al., 2011).

The main advantage of this configuration is that other types of interferometers (such as Fabry-Perot) can be inserted in the probe end with a reference provided by the Hi-Bi fiber section.

The second configuration proposed (Figure 3 b)) is more attractive in general. Here, the Hi-Bi fiber section is spliced to one output port of the second directional coupler. This means that the configuration can be used for remote sensing using only one fiber in the sensing zone. There is a section of an end mirrored Hi-Bi fiber at the second coupler output port, that will increase the returned optical power. It is clear that with this structure it is possible to have the sensing advantages of a FLM without the need of locating geometrically the sensing head inside the loop. A feature that increases the flexibility of the configuration shown in Figure 3 b) is the fact that the polarization controller is only needed to adjust the interferometric visibility. This configuration was compared with the conventional Hi-Bi FLM and presents similar sensitivity when the strain is applied (Frazao et al., 2011).

The last configuration (Figure 3 c)) was characterized as an interrogation setup for strain measurement. In this case, two Hi-Bi fiber sections are used as sensor and reference. The setup is formed by two simple couplers with low insertion loss, where the two output ports of the first coupler are spliced to the two input ports of the second one. In one of the splices, a Hi-Bi fiber section (reference) is inserted, while in the other arm a polarization controller is placed, in order to optimize the spectral response of the fringe pattern. The sensor is located at the output port of the second coupler and consists of the same type of Hi-Bi fiber. With the insertion of two Hi-Bi fiber sections, the configuration will have two combined unbalanced Sagnac interferometers. The Hi-Bi fiber used has a length of 1.34 m and 0.34 m for the reference and sensor signal, respectively. A signal processing based on the phase-shift quadrature of the two peaks power response of the reference signal was done in real time.

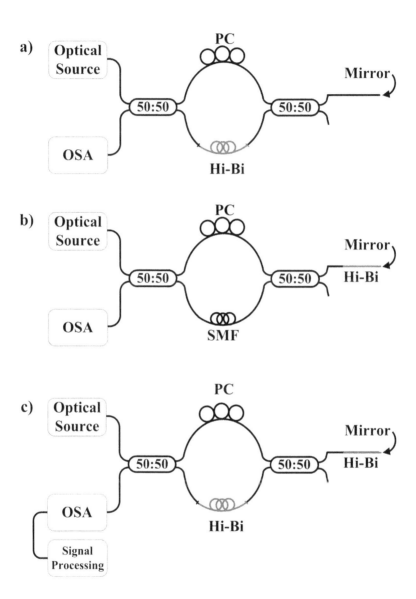

Fig. 3. Schematic setup for the: a) first; b) second; and c) third new Hi-Bi FLM configurations.

The sensitivity achieved in strain measurements is 16 mrad/με (Figure 4). The resolution of the system is 1.9 με, for a strain step variation of 29 με.

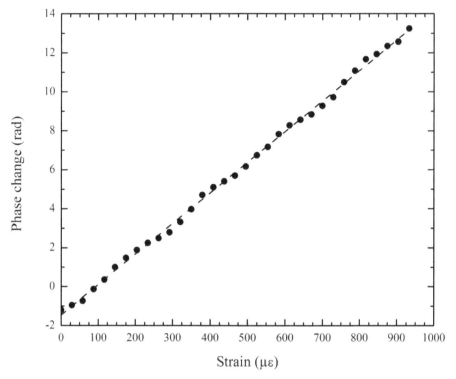

Fig. 4. Phase change versus strain.

4.2 Torsion sensor using a Sagnac interferometer

A temperature and strain independent torsion sensor is presented. In this case, the conventional optical coupler is replaced by Hi-Bi fiber coupler and the sensing head consists in a section of standard single mode fiber.

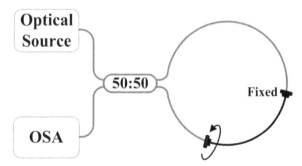

Fig. 5. Experimental setup of the torsion sensor based on Hi-Bi Sagnac loop interferometer.

Figure 5 presents the experimental setup of a Hi-Bi Sagnac interferometer sensor. Instead of using a regular fiber coupler, a 3 dB commercial Hi-Bi PANDA fiber coupler is used. This coupler will guarantee that the light will travel through the output arms with a constant polarization. A section of SMF is inserted in order to close the loop.

The interferometer is illuminated with a broadband optical source, and the transmission spectral response of the sensing head is measured with an optical spectrum analyzer (OSA).

The ports length difference of the Hi-Bi coupler will set the fringes pattern, since it introduces a phase accumulation to the travelling wave. The sensing head is located in the SMF region, and torsion measurements can be done by fixing one side of SMF section and twisting the other side, according to Figure 5. As the torsion is applied, an amplitude variation of the spectrum between maximum and minimum values is verified, which induces a change on the visibility of the fringes. Even though the amplitude changes with the torsion angle, total destructive interference is never observed, due to the existence of a beat between the two interferometers. The sensor response is periodically modulated, as the twist is applied. The sensitivity depends on the sensing head length, bigger sensing heads are more sensitive. For a 0.26 m long sensing head, sensitivities of 0.045 degree^{-1} and -0.051 degree^{-1} were obtained, when left and right twisting angles were respectively applied (Figure 6). The slight difference between the negative sensitivities can be due to the coating effect.

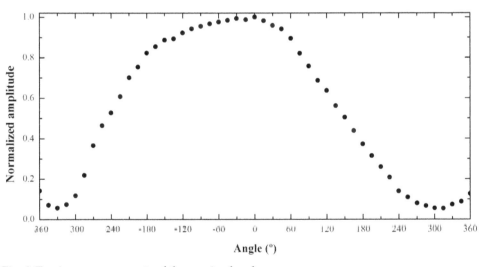

Fig. 6. Torsion measurements of the sensing head.

For temperature measurements, a sensing head with a length of 0.26 m is placed in a tubular oven, which enables the temperature to be set with an error smaller than 0.1 °C. Regarding strain characterization, the sensing head is attached to a translation stage with a resolution of 1 μm. The sensitivities obtained are 2×10^{-3} dB/°C and 7.9×10^{-9} dB/με, respectively. Due to the low sensitivities obtained, it can be stated that this sensing head is insensitive to these measurands. These results are expected because temperature and strain do not change the polarization states of the two counter propagating waves in the SMF region.

The main distinctive feature of this setup, when compared with a conventional Hi-Bi fiber loop mirror, is the polarization controller suppression.

5. Conclusion

The first optical fiber highly-birefringent Sagnac interferometer was presented in 1997 (Starodumov et al., 1997). This structure has been consistently used in the ambit of conception of new devices and systems in several areas related to the optical fiber technologies, namely in the optical fiber sensors context. By combining optical devices, it was possible to discriminate physical parameters, such as deformation and temperature. With the emerging of new fibers, like the Photonic Crystal Fibers, it is possible to measure one physical parameter with insensitivity to temperature. At last, with the development of new fiber loop mirror designs, it is possible to interrogate and measure the physical parameters with low cost and easily multiplexing with several sensing heads in series. All works recently published demonstrated the interferometric topology versatility of the Sagnac interferometer in the sensing domain. It must be pointed out that it shall be possible to obtain much better resolutions than the ones indicated if adequate signal processing techniques are applied, that allow reading the interferometer phase.

6. References

Andre, R. M., Marques, M. B., Roy, P. & Frazao, O. (2010). Fiber Loop Mirror Using a Small Core Microstructured Fiber for Strain and Temperature Discrimination. *IEEE Photonics Technology Letters*, vol. 22, No. 15, (Aug 1, 2010), pp. 1120-1122, ISSN 1041-1135

Campbell, M., Zheng, G., Holmes-Smith, A. S. & Wallace, P. A. (1999). A frequency-modulated continuous wave birefringent fibre-optic strain sensor based on a Sagnac ring configuration. *Measurement Science & Technology*, vol. 10, No. 3, (Mar, 1999), pp. 218-224, ISSN 0957-0233

Campbell, M., Zheng, G., Wallace, P. A. & Holmessmith, A. S. (1997). A distributed frequency modulation continuous wave fiber stress sensor based on a birefringent Sagnac ring configuration. *Optical Review*, vol. 4, No. 1A, (Jan-Feb, 1997), pp. 114-116, ISSN 1340-6000

Delarosa, E., Zenteno, L. A., Starodumov, A. N. & Monzon, D. (1997). All-fiber absolute temperature sensor using an unbalanced high-birefringence Sagnac loop. *Optics Letters*, vol. 22, No. 7, (Apr 1, 1997), pp. 481-483, ISSN 0146-9592

Dong, B., Hao, J. Z., Liaw, C. Y. & Xu, Z. W. (2011). Cladding-Mode Resonance in Polarization-Maintaining Photonic-Crystal-Fiber-Based Sagnac Interferometer and Its Application for Fiber Sensor. *Journal of Lightwave Technology*, vol. 29, No. 12, (Jun 15, 2011), pp. 1759-1763, ISSN 0733-8724

Dong, B., Zhao, Q. D., Feng, L. H., Guo, T., Xue, L. F., Li, S. H. & Gu, H. (2006). Liquid-level sensor with a high-birefringence-fiber loop mirror. *Applied Optics*, vol. 45, No. 30, (Oct 20, 2006), pp. 7767-7771, ISSN 0003-6935

Dong, X. Y., Tam, H. Y. & Shum, P. (2007). Temperature-insensitive strain sensor with polarization-maintaining photonic crystal fiber based Sagnac interferometer. *Applied Physics Letters*, vol. 90, No. 15, (Apr 9, 2007), pp. -, ISSN 0003-6951

Fang, X. J. & Claus, R. O. (1995). Polarization-Independent All-Fiber Wavelength-Division Multiplexer Based on a Sagnac Interferometer. *Optics Letters*, vol. 20, No. 20, (Oct 15, 1995), pp. 2146-2148, ISSN 0146-9592

Ferreira, M. S., Baptista, J. M., Roy, P., Jamier, R., Fevrier, S. & Frazao, O. (2011). Highly birefringent photonic bandgap Bragg fiber loop mirror for simultaneous measurement of strain and temperature. *Optics Letters*, vol. 36, No. 6, (Mar 15, 2011), pp. 993-995, ISSN 0146-9592

Frazao, O., Baptista, J. M. & Santos, J. L. (2007a). Recent advances in high-birefringence fiber loop mirror sensors. *Sensors*, vol. 7, No. 11, (Nov, 2007a), pp. 2970-2983, ISSN 1424-8220

Frazao, O., Baptista, J. M. & Santos, J. L. (2007b). Temperature-independent strain sensor based on a Hi-Bi photonic crystal fiber loop mirror. *IEEE Sensors Journal*, vol. 7, No. 9-10, (Sep-Oct, 2007b), pp. 1453-1455, ISSN 1530-437X

Frazao, O., Egypto, D., Bittencourt, L. A., Giraldi, M. T. M. R. & Marques, M. B. (2008a). Temperature sensor using hi-bi erbium-doped fiber loop mirror. *Microwave and Optical Technology Letters*, vol. 50, No. 12, (Dec, 2008a), pp. 3152-3154, ISSN 0895-2477

Frazao, O., Guerreiro, A., Santos, J. L. & Baptista, J. M. (2007c). Birefringence monitoring of a Hi-Bi fibre under chemical etching through a fibre loop mirror. *Measurement Science & Technology*, vol. 18, No. 12, (Dec, 2007c), pp. N81-N83, ISSN 0957-0233

Frazao, O., Marques, B. V., Jorge, P., Baptista, J. M. & Santos, J. L. (2008b). High birefringence D-type fibre loop mirror used as refractometer. *Sensors and Actuators B-Chemical*, vol. 135, No. 1, (Dec 10, 2008b), pp. 108-111, ISSN 0925-4005

Frazao, O., Marques, L. M. & Baptista, J. M. (2006a). Fibre Bragg grating interrogation based on high-birefringence fibre loop mirror for strain-temperature discrimination. *Microwave and Optical Technology Letters*, vol. 48, No. 11, (Nov, 2006a), pp. 2326-2328, ISSN 0895-2477

Frazao, O., Marques, L. M., Santos, S., Baptista, J. M. & Santos, J. L. (2006b). Simultaneous measurement for strain and temperature based on a long-period grating combined with a high-birefringence fiber loop mirror. *IEEE Photonics Technology Letters*, vol. 18, No. 21-24, (Nov-Dec, 2006b), pp. 2407-2409, ISSN 1041-1135

Frazao, O., Santos, J. L. & Baptista, J. M. (2007d). Strain and temperature discrimination using IF concatenated high-birefringence fiber loop mirrors. *IEEE Photonics Technology Letters*, vol. 19, No. 13-16, (Jul-Aug, 2007d), pp. 1260-1262, ISSN 1041-1135

Frazao, O., Silva, R. M., Kobelke, J. & Schuster, K. (2010). Temperature- and strain-independent torsion sensor using a fiber loop mirror based on suspended twin-core fiber. *Optics Letters*, vol. 35, No. 16, (Aug 15, 2010), pp. 2777-2779, ISSN 0146-9592

Frazao, O., Silva, R. M. & Santos, J. L. (2011). High-Birefringent Fiber Loop Mirror Sensors With an Output Port Probe. *IEEE Photonics Technology Letters*, vol. 23, No. 2, (Jan 15, 2011), pp. 103-105, ISSN 1041-1135

Frazao, O., Silva, S. O., Baptista, J. M., Santos, J. L., Statkiewicz-Barabach, G., Urbanczyk, W. & Wojcik, J. (2008c). Simultaneous measurement of multiparameters using a Sagnac interferometer with polarization maintaining side-hole fiber. *Applied Optics*, vol. 47, No. 27, (Sep 20, 2008c), pp. 4841-4848, ISSN 0003-6935

Fu, H. Y., Tam, H. Y., Shao, L. Y., Dong, X. Y., Wai, P. K. A., Lu, C. & Khijwania, S. K. (2008). Pressure sensor realized with polarization-maintaining photonic crystal fiber-based Sagnac interferometer. *Applied Optics*, vol. 47, No. 15, (May 20, 2008), pp. 2835-2839, ISSN 0003-6935

Fu, H. Y., Wong, A. C. L., Childs, P. A., Tam, H. Y., Liao, Y. B., Lu, C. & Wai, P. K. A. (2009). Multiplexing of polarization-maintaining photonic crystal fiber based Sagnac interferometric sensors. *Optics Express*, vol. 17, No. 21, (Oct 21, 2009), pp. 18501-18512, ISSN 1094-4087

Gong, H. P., Chan, C. C., Zu, P., Chen, L. H. & Dong, X. Y. (2010). Curvature measurement by using low-birefringence photonic crystal fiber based Sagnac loop. *Optics Communications*, vol. 283, No. 16, (Aug 15, 2010), pp. 3142-3144, ISSN 0030-4018

Han, Y. G., Chung, Y. & Lee, S. B. (2009). Discrimination of strain and temperature based on a polarization-maintaining photonic crystal fiber incorporating an erbium-doped fiber. *Optics Communications*, vol. 282, No. 11, (Jun 1, 2009), pp. 2161-2164, ISSN 0030-4018

Hwang, K. J., Kim, G. H., Lim, S. D., Lee, K., Park, J. W. & Lee, S. B. (2011). A Novel Birefringent Photonic Crystal Fiber and Its Application to Curvature Measurement. *Japanese Journal of Applied Physics*, vol. 50, No. 3, (Mar, 2011), pp. -, ISSN 0021-4922

Jin, Y. X., Chan, C. C., Zhang, Y. F., Dong, X. Y. & Zu, P. (2010). Temperature sensor based on a pressure-induced birefringent single-mode fiber loop mirror. *Measurement Science & Technology*, vol. 21, No. 6, (Jun, 2010), pp. -, ISSN 0957-0233

Kang, J. A., Dong, X. Y., Zhao, C. L., Qian, W. W. & Li, M. C. (2011). Simultaneous measurement of strain and temperature with a long-period fiber grating inscribed Sagnac interferometer. *Optics Communications*, vol. 284, No. 8, (Apr 15, 2011), pp. 2145-2148, ISSN 0030-4018

Kim, B. H., Lee, S. H., Lin, A. X., Lee, C. L., Lee, J. & Han, W. T. (2009). Large temperature sensitivity of Sagnac loop interferometer based on the birefringent holey fiber filled with metal indium. *Optics Express*, vol. 17, No. 3, (Feb 2, 2009), pp. 1789-1794, ISSN 1094-4087

Kim, D. H. & Kang, J. U. (2004). Sagnac loop interferometer based on polarization maintaining photonic crystal fiber with reduced temperature sensitivity. *Optics Express*, vol. 12, No. 19, (Sep 20, 2004), pp. 4490-4495, ISSN 1094-4087

Kim, G., Cho, T. Y., Hwang, K., Lee, K., Lee, K. S., Han, Y. G. & Lee, S. B. (2009). Strain and temperature sensitivities of an elliptical hollow-core photonic bandgap fiber based on Sagnac interferometer. *Optics Express*, vol. 17, No. 4, (Feb 16, 2009), pp. 2481-2486, ISSN 1094-4087

Kim, H. M., Kim, T. H., Kim, B. & Chung, Y. (2010). Temperature-Insensitive Torsion Sensor With Enhanced Sensitivity by Use of a Highly Birefringent Photonic Crystal Fiber. *IEEE Photonics Technology Letters*, vol. 22, No. 20, (Oct 15, 2010), pp. 1539-1541, ISSN 1041-1135

Lee, T. H., Lee, H. D., Kim, C. S. & Jeong, M. Y. (2010). Characterization of variable reflectivity of a polarization-maintaining fiber Sagnac mirror for long-distance remote fiber Bragg gratings cavity sensors. *Measurement Science & Technology*, vol. 21, No. 11, (Nov, 2010), pp. -, ISSN 0957-0233

Liu, Y., Liu, B., Feng, X. H., Zhang, W. G., Zhou, G., Yuan, S. Z., Kai, G. Y. & Dong, X. Y. (2005). High-birefringence fiber loop mirrors and their applications as sensors. *Applied Optics*, vol. 44, No. 12, (Apr 20, 2005), pp. 2382-2390, ISSN 0003-6935

Marques, B. V., Frazao, O., Mendonca, S., Perez, J., Marques, M. B., Santos, S. F. & Baptista, J. M. (2008). Optical current sensor based on metal coated Hi-Bi fiber loop mirror. *Microwave and Optical Technology Letters*, vol. 50, No. 3, (Mar, 2008), pp. 780-782, ISSN 0895-2477

Mortimore, D. B. (1988). Fiber Loop Reflectors. *Journal of Lightwave Technology*, vol. 6, No. 7, (Jul, 1988), pp. 1217-1224, ISSN 0733-8724

Qian, W. W., Zhao, C. L., Dong, X. Y. & Jin, W. (2010). Intensity measurement based temperature-independent strain sensor using a highly birefringent photonic crystal fiber loop mirror. *Optics Communications*, vol. 283, No. 24, (Dec 15, 2010), pp. 5250-5254, ISSN 0030-4018

Qian, W. W., Zhao, C. L., He, S. L., Dong, X. Y., Zhang, S. Q., Zhang, Z. X., Jin, S. Z., Guo, J. T. & Wei, H. F. (2011). High-sensitivity temperature sensor based on an alcohol-filled photonic crystal fiber loop mirror. *Optics Letters*, vol. 36, No. 9, (May 1, 2011), pp. 1548-1550, ISSN 0146-9592

Silva, R. M., Layeghi, A., Zibaii, M. I., Santos, J. L. & Frazão, O. (2011). Theoretical and Experimental Results of High Birefringent Fiber Loop Mirror with an Output Port Probe. *Journal of Lightwave Technology*, vol. -, No. -, 2011), pp. -, -(Submitted)

Starodumov, A. N., Zenteno, L. A., Monzon, D. & Delarosa, E. (1997). Fiber Sagnac interferometer temperature sensor. *Applied Physics Letters*, vol. 70, No. 1, (Jan 6, 1997), pp. 19-21, ISSN 0003-6951

Urquhart, P. (1989). Fiber Lasers with Loop Reflectors. *Applied Optics*, vol. 28, No. 17, (Sep 1, 1989), pp. 3759-3767, ISSN 0740-3224

Yang, X. F., Zhao, C. L., Peng, Q. Z., Zhou, X. Q. & Lu, C. (2005). FBG sensor interrogation with high temperature insensitivity by using a HiBi-PCF Sagnac loop filter. *Optics Communications*, vol. 250, No. 1-3, (Jun 1, 2005), pp. 63-68, ISSN 0030-4018

Zhang, H., Liu, B., Wang, Z., Luo, J. H., Wang, S. X., Jia, C. H. & Ma, X. R. (2010). Temperature-insensitive displacement sensor based on high-birefringence photonic crystal fiber loop mirror. *Optica Applicata*, vol. 40, No. 1, 2010), pp. 209-217, ISSN 0078-5466

Zhao, C. L., Yang, X. F., Lu, C., Jin, W. & Demokan, M. S. (2004). Temperature-insensitive interferometer using a highly birefringent photonic crystal fiber loop mirror. *IEEE Photonics Technology Letters*, vol. 16, No. 11, (Nov, 2004), pp. 2535-2537, ISSN 1041-1135

Zhou, D. P., Wei, L., Liu, W. K. & Lit, J. W. Y. (2008). Simultaneous measurement of strain and temperature based on a fiber Bragg grating combined with a high-birefringence fiber loop mirror. *Optics Communications*, vol. 281, No. 18, (Sep 15, 2008), pp. 4640-4643, ISSN 0030-4018

Zibaii, M. I., Latifi, H., Frazão, O. & Jorge, P. a. S. (2011). Controlling the Sensitivity of Refractive Index Measurement using a Tapered Fiber Loop Mirror. *IEEE Photonics Technology Letters*, (2011), (Accepted)

Zu, P., Chan, C. C., Jin, Y. X., Zhang, Y. F. & Dong, X. Y. (2011a). Fabrication of a temperature-insensitive transverse mechanical load sensor by using a photonic crystal fiber-based Sagnac loop. *Measurement Science & Technology*, vol. 22, No. 2, (Feb, 2011a), ISSN 0957-0233

Zu, P., Chan, C. C., Siang, L. W., Jin, Y. X., Zhang, Y. F., Fen, L. H., Chen, L. H. & Dong, X. Y. (2011b). Magneto-optic fiber Sagnac modulator based on magnetic fluids. *Optics Letters*, vol. 36, No. 8, (Apr 15, 2011b), pp. 1425-1427, ISSN 0146-9592

5

Sensing Applications for Plastic Optical Fibres in Civil Engineering

Kevin S. C. Kuang
National University of Singapore,
Department of Civil and Environmental Engineering
Singapore

1. Introduction

Monitoring of the performance and integrity of structures using sensors attached or embedded to them has been of great interest in view of the possibility of predicting potential structural failures, optimising the operational efficiency of the structure, extending the useful lifespan of the host structure as well as tracking the propagation of cracks, amongst other benefits. Indeed, the field of structural health monitoring has received considerable attention and investments from private and public institutions including academia, professional institutes, industrial research laboratories, government agencies as well as the industry. A wide range of companies involved in the designing, fabrication, integration, usage, servicing and maintenance of structural components, assemblies and systems have a common need to monitor, understand and manage the performance and integrity of the structures in their possession. These companies hail from industries such as the offshore, oil-and-gas, power generation, building construction, marine, aerospace, civil and environmental.

Monitoring the performance and integrity of a host structure is often translated into the measurement of key parameters or health signatures of the structure as indicator of the integrity and level of performance of the host. Common parameters of interest include strain, load, deflection, displacement, pH-level, temperature, pressure, vibration frequency, liquid level, detection of cracks and monitoring of their rates of propagation, moisture, detection of events such as impact, overload, fracture and others. Several review papers are available in the open literature outlining various examples of optical fibre sensor applications [1-3].

In recent years, optical fibre sensor technology has gained significant attraction as the sensor of choice in many structural health-monitoring applications. Several optical fibre sensing techniques have been commercialised whilst others are being developed. These sensors in general, offer several distinctive benefits and advantages over conventional sensors - these include their insensitivity to electromagnetic and microwave radiation (occurring near power-generating equipment), being spark-free (hence suitable for use in areas where risk of explosion is a concern), the fact that they are intrinsically safe (inert and passive), non-conductive, lightweight and potentially suitable for embedding into structures such as fibre composite laminates during the fabrication stage.

There are several unique features associated with optical fibre sensors. Certain types of optical fibre sensors (e.g. fibre Bragg gratings) possess the possibility of having multiple sensing regions in a single strand of fibre - this capability where multiple sensors could be interrogated simultaneously in quasi-real time using a single channel is a significant advantage over conventional sensors such as strain gauges where a single channel monitors only a single sensor. In addition, recent advancement in the development of interrogators based on optical time-domain reflectometry (OTDR) principle has opened up the possibility of performing distributed monitoring of strain and temperature along a single fibre sensor. Also, although optical fibre sensing techniques, in general, are known to be capable of obtaining high measurement resolution (e.g. sub-microstrain measurement is possible with fibre Bragg grating sensor), their ability to monitor large strains (e.g. 40% strain or more) has also been reported [4]. In applications requiring the measurement of parameters that modulate over time (e.g. as in fluctuating liquid level [5]), optical fibre sensors offer a range of dynamic characteristics or responses which could be tailored to suit specific needs.

There are two main types of optical fibres based on the material from which they are made and these can be classified as either glass-type optical fibres (GOFs) or polymer-type optical fibres (POFs). Within each class of fibres, a variety of materials have been used, developed, doped and purified to achieve low transmission attenuation. Historically, optical fibre sensors were developed using GOFs due to their availability, excellent transmission properties, and other optical characteristics which make them amendable to sensor fabrication – e.g. the ease of changing the refractive-index of a section of a specially-doped glass fibres via light radiation has led to the development of fibre Bragg grating sensors. In recent years, however, interest in the use of POFs for short distance networking (e.g. local area networking), significant improvements in transmission properties, lower cost and various practical advantages of POFs over GOFs have spurred intense research and development in sensor technologies based on polymer fibres. Whilst many of the sensing methodologies developed based on GOFs could also be applied to POFs, the unique features and advantages of POFs opened up new possibilities and opportunities for its applications in the field of civil engineering.

Plastic optical fibres usually come with core sizes ranging from diameters of 0.25mm to 1mm although fibres of larger diameters could also be purchased - a search on the Internet will reveal many suppliers of POFs. The core of the fibre could be made from poly-methymethacyrlate (PMMA), polycarbonate (PC), polystyrene (PS) and cyclic-transparent optical polymer (CYTOP), this being a recent addition to the list. Although the transmission attenuation of POF based on PMMA (160 dB/km) is inferior to the CYTOP-type fibre, it is the most readily available POF. Being inexpensive, many cost-effective sensors have been developed based on PMMA step-index fibre for a variety of applications in civil engineering.

POF sensors have attracted increasing interest for use in health-monitoring of civil structures due to a number of advantages over their silica counterpart. These include their lower cost (the fibre itself and the associated accessories and hardware), less susceptible to fracture and flexibility compared to bare GOFs. They also offer ease of termination (i.e. cleaving can be done with a razor blade and connected easily to the ferrule), safe disposability and ease of handling during sensor fabrication and installation to the host structure. Compared to GOFs of the same diameter, POFs are lighter, cheaper and less prone to flexural damage. In addition, POF has an elastic limit of an order higher than GOF [6,7] and it has been shown that standard step-index PMMA POF can be integrated into geotextile materials to measure strains up to 40%. When it is desirable to embed sensors within concrete structures, POF sensors offer a possible solution since the extremely alkaline (pH 12) environment of the concrete mixture is known to be corrosive to standard glass fibres [8]. Furthermore, the presence of moisture can weaken the glass core and accelerates crack growth in the fibre. Although a polymer coating may be applied in order to protect the glass fibre from the corrosive environment, this will incur additional cost and change the sensing properties of the sensor.

The brief introduction above serves as the background information to the following examples of applications encountered in civil engineering. Initially, examples of POF sensors based on intensiometric principles will be highlighted, followed by techniques based on OTDR. This chapter will focus on these two particular techniques in view of the author's experience in working with these in many projects over the last ten years. The principle of operation of each technique will be outlined in each category prior to a showcase of some examples to assist the reader in appreciating the subject matter.

The aim of this chapter is to present to the reader the potential of POF sensing technique as an attractive option for various applications in civil engineering and is not intended to be a comprehensive review of the various studies published in the literature in the area of POF sensor. Through illustrative examples of applications, it will be shown that POF sensors are ideal candidates in numerous civil engineering projects, either complementing or replacing conventional sensors. In some cases, POF sensors will be shown to offer a solution not possible with existing commercial sensors in the same class. With further development aimed at exploiting the unique features of POF-based sensors they may well prevail as the sensor of choice in many civil engineering applications where their cost-effectiveness, reliability, simplicity in design and ability to perform under demanding conditions are the key considerations in the sensor selection process.

2. POF sensors for civil engineering applications

To date, there are several sensing methodologies developed for the POF. These techniques differ in the way which the optical signal (which contains information on the measured quantity) is interrogated. Depending on which optical characteristic was monitored, the technique could be classified as intensity-based, interference-based, polarisation-based, wavelength-based or OTDR-based.

Of the various types listed, intensity-based sensing technique represents one of the earliest and perhaps the most direct and inexpensive way of interrogating optical fibre sensors for structural health monitoring applications. The sensing principle is straightforward and relies on the monitoring of the intensity level of the optical signal as it modulates in response to the measured quantity. Although intensity fluctuation in the optical signal due to possible power variation at the source as well as influence of external parameters unrelated to the measured quantity (e.g. micro- and macro-bending along the fibre length) can occur, these could be overcome. For example, with the availability of stable and inexpensive light sources and low bend-sensitivity fibres, the intensity-based approach offers excellent technical and commercial prospects for high volume, large-scale applications from a cost-effectiveness point of view.

Intensity-based technique is also well-suited for vibration measurement since absolute strain values are not required for this quantity – provided that the oscillation is within the sensitivity of the sensor. In recent years, POF-based accelerometers have been developed with vibration monitoring capability and it has been demonstrated that these optical fibre-based vibration sensors compare well with conventional capacitive accelerometers for vibration up to 1 kHz.

In addition, intensity-based technique can be integrated to wireless modules with ease for monitoring of static and dynamic loads. This can be carried out at a low-cost and with ease in view of the availability of miniaturised solid-state devices which could be used as light sources and detectors for the POF sensor. An example of this concept has been demonstrated for a flood monitoring application [5]. These features highlight the potential of intensity-based technique for field deployment where their sensing capabilities could be exploited whilst achieving cost-effectiveness. Instead of the common approach which is to use an optical fibre technology based on best-in-class for their measurement needs (e.g. high strain resolution of fibre Bragg grating sensors, which often entails a high up-front cost) potential users need to be clear in their requirements and be open to evaluate intensity-based optical techniques which to provide a more cost-effective solution.

In this section, the principle of operation of each type of interrogation method covered in this chapter will be outlined accordingly. The two main types considered in this chapter are (1) Intensity-based sensing and (2) OTDR-based sensing.

2.1 Intensity-based POF sensors

2.1.1 Operating principles and background of intensity-based POF sensors

The ease of monitoring the intensity of the optical signal in POF naturally leads to the investigation of their potential as intensity-based sensors. Light sources are easily available and inexpensive and could come in the form of high-intensity light-emitting diodes (LEDs), typically those operating at the red region (centred at 625 nm) region of the visible light spectrum. Light detectors are also widely availabily and those that are commonly used in association with POF are light-dependant resistors (LDRs) or solid-state photodetectors. The interrogation technique is straightforward and involves monitoring the modulation of the light intensity as a function of an external perturbation of interest which include strain, displacement, load, pressure and others. Indeed the simplicity in design of the sensor as well as the low cost of the light source and detector associated with intensity-based technique have led to numerous applications not just for civil engineering but also for many other fields where detection and monitoring are required.

In general there are two categories of intensity-based POF sensors and these are classified as either (a) intrinsic or (b) extrinsic sensors. In the former category, the modulation of the intensity of the optical signal in the fibre is a result of the physical change in the optical fibre due to an external perturbation, such as load, strain or pressure acting on the fibre itself. This phenomenon can be observed when a sensitised region of the POF is bent resulting in the modulation of the light intensity. In the latter category of sensor, the modulation of the optical signal in the fibre is a result of a change in the physical-optic state of a transducer attached to the fibre resulting in the modulation of light intensity.

The optical fibre in this case serves as a light conduit and is not physically affected by the external perturbation itself. An example of this effect is when the light intensity modulates as a result of a change in the distance between two cleaved fibres resulting from an external perturbation.

2.1.2 Applications of intensity-based POF sensors

Intensity-based POF sensors have been demonstrated for a number of structural applications ranging from monitoring of flexural loading to detection of cracks in concrete specimens. In one of their earlier works, Kuang et al. [9] investigated the use of such type of POF sensor to monitor the mechanical response of a number of plastic specimens. The POF utilised in their studies was a step-index 1mm diameter PMMA POF supplied by Mitsubishi Rayon Co. Ltd. The flexural sensitivity of the fibre used (ESKA CK40) was enhanced by removing the cladding layer of a segment of the POF over a pre-determined length, the aim being to promote light loss in this region due to the reduction in the number of modes undergoing total internal reflection when the fibre was bent. In their studies, the sensitised regions reported range from 70mm for smaller specimens to 300mm for larger specimens were shown to be appropriate although it may be postulated that the length of choice will depend largely on the sensitivity required and the size of the host structure of the actual application. The study also demonstrated that the POF sensor based on the removal of cladding layer possess directional sensitivity in that highest flexural sensitivity could be achieved when the plane of cut resulting from the removal procedure was perpendicular to the direction of flex. The flexural strain-optic coefficient was reported to be approximately $1.8 \times 10^{-5}/\mu\varepsilon$, implying that at 1% flexural strain, a normalised loss of 18% in light intensity could be expected.

Since the removal of the cladding layer was done manually by scrapping the POF with a razor blade, the light intensity transmitted through to the fibre end would have decreased considerable with each subsequent scrapping process since with each cut, more of the POF core was removed. To achieve consistency in the amount of material removed, the procedure should be done with the POF placed in grooves machined to specific depths. The scrapping of the POF material would then be carried out in steps of decreasing groove depth as desired. A photograph showing the cross-section of the sensitised POF is given in Figure 1.

To test the response of the POF sensor under flexural load and tensile load, the authors mounted the POF onto its respective specimen using an acrylate-based adhesive. For the flexural load condition, a section of a straight fibre was aligned along the longitudinal axis of the specimen and attached to the bottom surface of the plastic specimen. The specimen was tested under a three-point bend set-up to evaluate the response of the sensor. To sensitise the POF to the flexural load, a segment of the fibre attached to the specimen was removed as described above. To test the POF response to tensile loading, the sensing region of the POF was curved slightly such that the fibre would tend to straighten out as the tensile load was applied.

Figure 2(a) shows the repeatability of the POF sensor response under a cyclic flexural loading, demonstrating the potential of the sensor to monitor the flexural displacement of the specimen reliably. The response of the POF sensor when tested in the tensile direction is

Fig. 1. Photograph showing the cross-sectional view of the sensitised POF where a segment of the POF (cladding and core) was removed (after [9]).

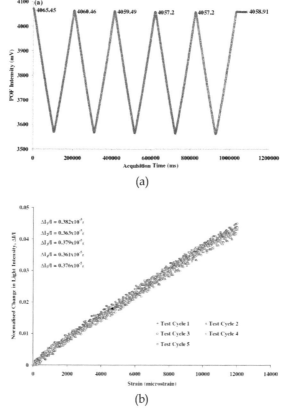

(a)

(b)

Fig. 2. Plot showing the response of the POF sensor during (a) cyclic flexural loading, (b) tensile loading (after [9]).

shown in Figure 2(b). Here the result also illustrates the high degree of repeatability of the response of the POF. It was highlighted that although the tests stopped at flexural and tensile strains of 0.7% and 1.2% respectively, the ability of the POF sensor design used in this study to measure strain values higher than that observed in the tests can be achieved. This may be supported by the fact that at those levels of strain observed in the study, the loss of intensity in each case did not show evidence of a plateau. In addition, studies conducted by other workers up to 45% [6,7] can be expected while other studies have shown that POF could endure stretching of more than 80% strain under certain conditions [10].

The performance of the POF was also investigated under dynamic loading conditions to assess their potential for vibration measurement, a technique frequently employed in health monitoring of civil engineering structures. Here, Kuang et al. [11] used the same POF sensor design described above in a set-up involving monitoring the natural frequencies of a series of increasingly damaged composite beams subjected to various degrees of damaged induced by low-velocity impact. The ability to detect the change in the frequency response of the host beam as a result of impact-induced damage is shown in Figure 3(a). In addition, the damping ratio of the cantilever set-up was also monitored. The ability of the sensor to monitor the damping ratio as a function of impact energy (inferring degree of damage) is shown in Figure 3(b). The changes in natural frequency and the decrease in the damping ratio were used to characterise the reduction in the post-impact flexural stiffnesses and the residual strengths of the composite beams with increasing level of impact damage. The study showed that the POF sensor was evidently able to detect a change in the damping ratio as small as 2.5%. The natural frequencies of the beams at various stages of damage based on the results of the POF sensor were found to compare well with the predicted theoretical values as well as the readings obtained from an electrical strain gauge. The authors also discovered in their study that impact tests on simply supported carbon-fibre reinforced beams also showed that the sensor was able to monitor out-of-plane deflections during the impact event. A laser Doppler velocimeter was also used in conjunction with a piezoelectric load cell as a means to validate the results of the POF sensor.

The ability to monitor the dynamic response of the host beam encouraged the authors to further apply the technique to monitor the morphing response of a nickel-titanium fibre- metal laminate [12]. A collocated electrical strain gauge provided a reference measurement to compare the response of the POF sensor. It was reported that following activation of the smart fibre metal laminate using a heat gun, the nickel-titanium deformed as predicted and the POF sensor was found to monitor the flexural response of the smart composite accurately. The response of the POF was compared to the electrical strain gauge and this is summarised in Figure 4 showing the high degree of similarity in the responses of both sensors.

This type of intensity-based POF sensor also found applications in crack detection and monitoring of deflection of civil structures. Kuang et al. [13] conducted a series of flexural tests on scaled-specimens where the POF sensor was attached to the bottom surface of the concrete beams and demonstrated in their study that the POF sensor design used were of sufficient sensitivity to detect the presence of hair-line cracks. Figure 5 shows photomicrographs of the size of the cracks in the concrete relative to the POF. The smallest crack width measured approximately 0.04 mm was successfully detected using the POF sensor. In order to improve the sensitivity of the POF to the beam deflection and crack initiation, a segment of the POF was

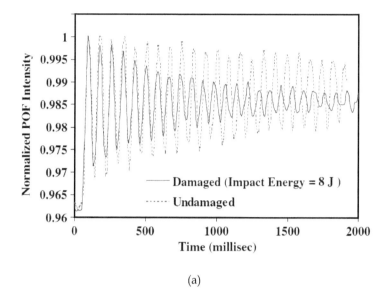

(a)

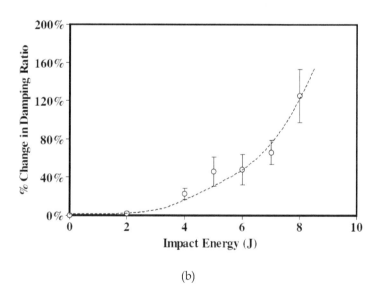

(b)

Fig. 3. Plots showing (a) a comparison of the dynamic response of undamaged and damaged specimen as measured by the POF sensor, (b) the ability of the POF sensor to monitor the change in the damping ratio response for an increasingly damaged specimen (after [11]).

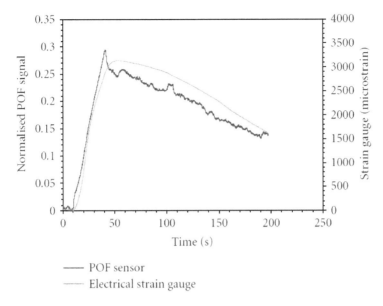

— POF sensor
— Electrical strain gauge

Fig. 4. Plots showing response of the POF in monitoring the fibre metal laminate upon activation of the nickel-titanium (after [12]).

Fig. 5. Photographs showing intersection of the crack and the POF sensor for the concrete specimens at difference degrees of crack width (after [13]).

removed as described earlier over a predetermined length (70 mm for a series of scaled un-reinforced concrete specimens and 300 mm for full-scale reinforced concrete specimens). For the scaled specimens, the authors showed that under three-point bend loading, the POF sensor was able to detect the presence of crack once it intersected the sensor and this observation was repeatable for the entire series of specimens as shown in Figure 6. Similarly, for the full-size specimens the test results affirmed the potential of the POF sensor to monitor the response of the concrete beam under load. Initial cracks in the concrete beam were successfully detected and the POF continued to monitor the post-crack vertical deflection through to finally detecting the failure cracks in the specimen.

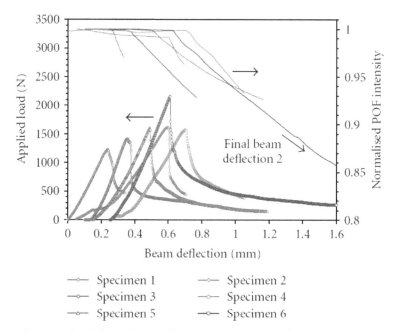

—○— Specimen 1   —○— Specimen 2
—○— Specimen 3   —○— Specimen 4
—○— Specimen 5   —○— Specimen 6

Fig. 6. Plots showing the ability of the POF sensors in detecting the crack for a series of six scaled concrete specimens (after [13]).

In a different area of application within the field of civil engineering, POF sensor has been demonstrated to have potential to be used as flood sensors. Monitoring of the water level in flood-prone areas could provide early warning of flash floods which devastated various regions in ASEAN (Association of the Southeast Asian Nations) in recent times, both in the urban and rural areas. The monitoring of liquid level is also an important activity in many industries and the special requirements to monitor a large number of multiple containers of volatile fluids demands the implementation of cost-effective and safe sensor systems. Other applications where liquid level monitoring is desirable include monitoring of the water level in catchment areas, monitoring of illegal dumping of trash and pollution in canals (which can be inferred by an unexpected rise of water level) and early warning of impending drought (when water levels drops to a minimum level). Information collected by a network of sensors could also be used to complement and update flood monitoring and prediction software models.

The intensity-based POF sensor proposed for flood monitoring by Kuang et al. [5] was simplified from earlier designs applied to glass-type fibres. In the simplified design, neither fusion-splicing nor heating were used in the sensor construction as the POF could be readily bent to shape without any risk of fracture. No prismatic element was used in the proposed design, thus rendering its construction significantly easier whilst exhibiting excellent optical response. It was argued that the simplicity in design of the proposed sensor could lead to an inexpensive, rugged, easy-to-use and effective optical fibre-based sensor for mass deployment in a flooding monitoring application. In addition, the authors demonstrated the ease of integrating the POF sensor to a commercial wireless mote, utilizing the on-board light dependent resistor to monitor the optical intensity transmitted through the fibre. Figure 7 shows the assembled unit consisting of the POF sensor and the wireless mote. The convenience and ease of integration the POF to off-the-shelve wireless motes would not be possible with other interrogation techniques, highlighting the unique advantage of intensity-based approach.

Fig. 7. Photograph of the wireless POF prototype and the MICA2DOT unit used in the flood monitoring experiment (after [5]).

The principle of operation of the POF sensor used in the flood monitoring study was based on the loss of total internal reflection or absorption of evanescence wave when the probe was immersed into the liquid medium. Before the sensitized probe was immersed into the liquid (i.e. in air which has a refractive index of ~1), the majority of the light ray will undergo total internal reflection and will be guided along the optical fibre to the output end of the fibre. However, when the probe was immersed into a liquid with a higher refractive index (e.g. water, which has a refractive index of 1.3), light traveling as cladding-mode will be absorbed by the liquid as the condition for total internal reflection could not be satisfied at the fibre–water interface. The intensity of the optical signal arriving at the output end of the fibre will correspondingly be reduced. It follows that when the probe was immersed in a liquid with higher refractive index, more light will be lost from the optical fibre into the liquid. This being a point sensing system, several such sensors may be deployed concurrently to achieve a multi-level discrete measurement of liquid level.

The response of the sensor was also tested and Figure 8 shows the repeatability of the liquid detection capability of the proposed sensor. From the sensor response, it is clear that it exhibits excellent repeatability showing no sign of hysteresis. At the circle marked A, corresponding to the moment the tip of the probe touched the surface of the liquid, the displacement transducer reading was initialized and set to zero. At that moment, the sensor signal decreased sharply as light escaped into the test liquid. When the probe was being drawn out, the sensor signal began increasing gradually starting from time B when the probe was beginning to leave the surface of the liquid. This occurred at approximately the same probe position corresponding to that of circle A. As a result of the surface tension of the liquid, a cusp formed at the tip of the probe as it past the zero position. The wetted area at the tip decreased gradually as the probe was being drawn out and this corresponds to the increasing signal reading as more light rays undergo total internal reflection. As the cusp broke at time corresponding to circle C, the sensor signal reaches maximum value at approximate probe position of 0.37mm. From the chart, the response of the sensor when withdrawn from the liquid lags behind the physical position of the probe as defined by the difference in the distance between positions A and C, this interval being approximately 0.37mm for the liquid used. Clearly, the value of this interval, which is a function of the surface tension, will vary accordingly with liquid used. However, level of accuracy better than 1mm may not be at all necessary for many applications in civil engineering such as flood monitoring or water level in reservoirs. This example provides further evidence of the cost-effectiveness of the intensity-based approach in meeting the sensing needs of an important area in civil engineering.

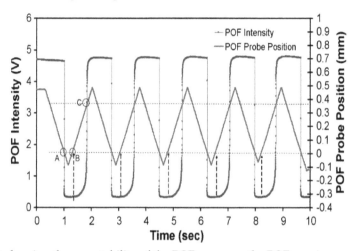

Fig. 8. Chart showing the repeatability of the POF sensor as the POF was immersed in and removed from water (after [5]).

Another significant design based on intensity sensing technique relies on the principle that a change in the gap between two cleaved POF surfaces resulting from an applied strain will leads to a corresponding chance in the light intensity of the sensor [14]. The operating principle of the POF sensor is straightforward- the sensor relies on the monitoring of the optical power transmitted through an air-gap between two cleaved optical fibre surfaces. The two fibres are aligned within a housing in which the fibre could slide smoothly. In this case, the housing itself is strain-free while the POFs are displaced relative to each other during the bending of the beam. It is important to note that the two fibres on either ends of the housing do not undergo any strain since they are free to slide relative to the housing. The gap changes in proportion to the applied strain leading to the modulation of the transmitted optical power.

Two versions of the sensor design are shown schematically in Figure 9. The authors showed that by adding a suitable opaque liquid, the strain sensitivity of the sensor could be improved significantly compared to one without the inclusion of the liquid. A comparison of the strain (or extension) sensitivity of the sensor is shown in Figure 10, highlighting the effect of different liquid opacity on its sensitivity. In the study, four POF sensors and an electrical strain gauge were attached to the bottom surface of a reinforced scaled concrete

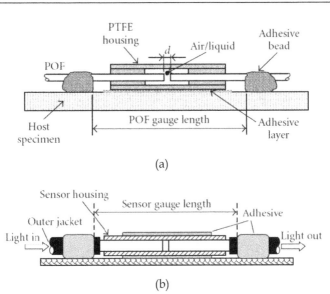

(a)

(b)

Fig. 9. Schematic showing the two versions of the POF sensor design based on changes in the gap between two cleaved fibres (a) (after [14]), (b)(after [4]).

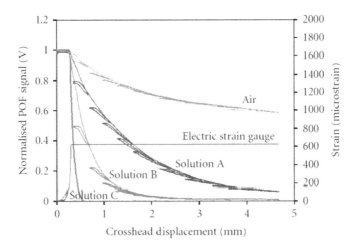

Fig. 10. Plot showing the different responses of the four POF sensors and the electrical strain gauge under a cyclic loading test (after [14]).

beam. The specimen was subjected to a flexural cyclic loading in a three-point bent set-up. When the failure load was reached, a crack developed across the beam which also damaged the electrical strain gauge. The four POF sensors were found to have survived and were able to continue monitoring the loading cycle as shown in Figure 11.

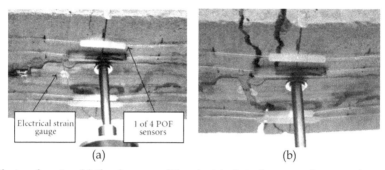

(a)   (b)

Fig. 11. Photos showing (a) the damage of the electrical strain gauge due to an intersecting crack (b) the relatively better survivability of the four POF sensors (no visible damage) even under severe crack (after [14]).

Building on the work conducted earlier, here the authors made further progress in demonstrating the ease of integrating the signal of the intensity-based POF sensor as a feedback to a control system to direct the amount of deflection in a smart fibre laminate beam [15]. Figure 12 shows a schematic of the set-up used in the experiment. Here, in order to activate the morphing of the smart beam, a thin-film heating source was attached to the beam. The POF sensor, attached to the opposite surface provided the feedback signal to the controller such that the appropriate amount of power was supplied to the heater to allow the smart beam to deflect to the desired position. The deflection of the beam was monitored continuously by the POF sensor and used as feedback to the controller to achieve the pre-set deflection position accurately. Figure 13 shows the POF sensor readings at three different beam deflections (A) to (C). The result illustrates that the POF-instrumented beam could be controlled accurately to within 3% of the desired deflection using the feedback data from the sensor with very little overshoot.

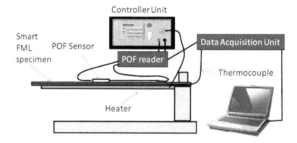

Fig. 12. Experimental set-up used to evaluate the potential use of the POF as feedback sensor to control amount of deflection of the smart composite beam (after [15]).

It was also of interest to gain an insight of the ability of the POF sensor to detect external disturbances acting on the beam and the ability of the control system to use the feedback data from the POF sensor to self-correct the amount of deflection of the smart beam. When a weight was added to the free end of the specimen after the smart beam had achieved steady-state, the POF immediately detected the change in the deflection of the beam which in turn resulted in a decrease in the POF sensor output triggering the controller to activate the

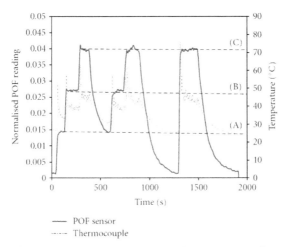

Fig. 13. Plot showing the response of the POF sensor demonstrating the potential of using the POF sensor to provide feedback signal achieve the amount of deflection desired in the smart composite beam (after [15]).

shape memory alloy via the heater. On heating to its activation temperature, the smart beam position was restored as evidenced by the steady state values of the POF sensor at the original controlled position. When the weight was removed, the position of the beam was again perturbed. Here the POF sensor triggered a corresponding series of action similar as before resulting in the beam settling to the controlled position as illustrated in the POF trace in Figure 14.

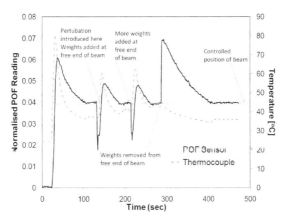

Fig. 14. Plot showing the self-correction capability of the control system based on the POF sensor output (after [15]).

The ability of the above POF sensor to monitor relatively large extension was further studied in a separate application [4, 16]. Here, the POF sensor based on Figure 9(b) was attached to two types of polypropylene geotextile specimens as shown in Figure 15, one being woven whilst the other being non-woven. A series of specimens, instrumented with

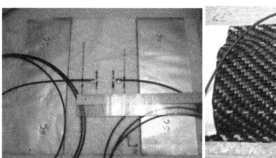

Fig. 15. Photographs showing the two types of POF-instrumented geotextile specimens (after [16]).

the POF sensor, were tested under quasi-static tensile load as well as being subjected to an acceleration field of 100g in a geotechnical centrifuge machine in a submerged condition.

In the quasi-static tensile test, the geotextile specimens were loaded to failure and the typical results are shown in Figure 16. The results show the ability of the POF sensor to monitor strains exceeding 35% for the non-woven specimens and approximately 20% for the woven specimens. The responses of the POF sensors in both cases compared very well with other reference measurement techniques. The POF sensor is expected to be able to monitor even higher strains since the fibre itself does not undergo any strain. It is noteworthy that the sensing principle relies on the changes in the distance of the gap between two fibres and the adhesive to attach the sensor to the host substrate was applied on the fibre, not on the housing itself, hence, neither the fibre nor the housing experience any significant strain, since the strain limit is a function of the gauge length (i.e. distance between the adhesives applied on both sides of the fibre) and the provision of sufficient fibre length within the housing is important such that at maximum extension, the fibre does not protrude from the housing.

In the centrifuge test, it was reported that the POF sensors survived the acceleration of 100g whilst submerged under water. Here, the instrumented geotextile tube was released from a hopper and dumped to the bottom of the centrifuge box to study the strain development of the geotextile tube throughout the whole of this process. A study of the sensors output showed that the POF sensor was able to monitor the strain experienced by the model geotextile tube reasonably well, following a similar trend and magnitude of measurement as obtained using the electrical strain gauge as illustrated in Figure 17. Towards the end of the test, it was noted that the two sensors gave different responses; the POF sensor indicated a sharp increase in strain while the strain gauge recorded a decrease. The resistance type strain gauge was deemed to have correctly recorded the decreasing strain trend. It is postulated that since the geotextile tube came to its resting position on the base of the strong box, without any external loading, the strain developed on the base centre portion of the geotextile would be insignificant, as registered by the strain gauge. On inspection of the POF sensor, it was noted that the fibre ends of POF sensor had protruded out of its housing which led to the sudden loss of intensity at model time 6 s as indicated in Figure 17. It was likely that the POF was tugged as the geotextile tube was being released resulting in the separation of the fibre from its housing. Although the POF sensor was damaged during the last phase of the test, the centrifuge experiment was encouraging, demonstrating the potential of the POF sensor for measuring large strains in geotextile tubes.

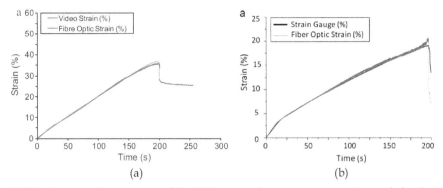

(a)   (b)

Fig. 16. Plot showing the response of the POF sensor during a quasi-static tensile loading for (a) a non-woven specimen, (b) a woven specimen (after [16]).

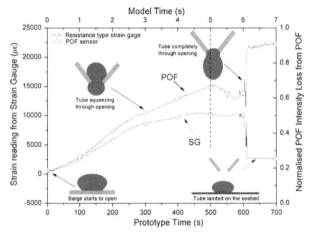

Fig. 17. Plot showing the excellent response of the POF sensor in the centrifuge test compared to the electrical strain gauge (after [16]).

2.2 OTDR-based POF sensors

OTDR-based POF sensors have received some attention recently partly in view of the recent advancement in the field of optical time-domain reflectrometry for POF. Developmental work in the area of the interrogator has resulted in the availability of high-resolution photon-counting systems allowing distributed monitoring along a single POF fibre. OTDR is a well-known technique in telecommunication for fault analysis and crack detection in silica fibres. The ability to detect the very low intensity light levels and high detection speed whilst achieving a high level of spatial resolution is not possible with standard OTDRs. The use of graded-index fibre in conjunction with v-OTDR also reduces the modal dispersion encountered in step-index PMMA fibres and hence graded index fibre is also used for monitoring of fibre over longer distances compared to step-index fibres. One of the most significant advantages of the OTDR technique is the possibility of distributed sensing, i.e. monitoring along the entire length of the fibre is in principle achievable.

2.2.1 Operating principles and background of OTDR-based POF sensors

OTDR sensing exploits the monitoring of the backscatter light in an optical fibre following the launched of a short optical pulse at one end of the fibre. The ability of the OTDR to detect the position of a defect along its length is based on the principle of time-of-flight. The time it takes for the pulse to be reflected back to the detector as backscatter signal is measured very precisely i.e. the backscatter signal is recorded as a function of time which is then converted to a distance measurement. Perturbations, such as strain or defects along the length of the fibre will result in either a peak Fresnel reflection or loss in the backscatter signal at the location of the perturbation. Specifically for POF use, a few companies have developed and commercialized v-OTDR to detect the very weak backscatter light signal due to the high attenuation and large dispersion in POFs. The photon-counting technology, based on an avalanche photodiode allows detection of very short optical pulses (~1 ns) necessary to resolve the location of defects accurately (to within 10 cm resolution).

An OTDR trace consists of a number of reflection peaks corresponding to specific physical junctions along the fibre such as splices, connectors, fibre terminations or defects. The trace is typically plotted on an intensity-distance graph where the location of the peaks along the x-axis refers to their positions of the event or physical entity (splices, connectors, defects etc.) along the fibre. Hence the position of a crack along the fibre will be shown as a reflection peak on the OTDR trace. In order to match the peaks in the OTDR trace to their respective event, the position of a physical reference along the fibre such as a connector, should be defined. Unknown positions of defects or cracks anywhere along the fibre can then be determined accurately with reference to the pre-defined position.

When used to detect damage in structures, the POF should clearly be sufficiently sensitive to detect the parameter of interest. Since detection of cracks along the fibre is an inherent capability of the OTDR-technique, a possible application of the technique is to apply it for crack detection in structures such as steel tubes. Other applications, which will not be included in this chapter, include large strain measurement in geotextile, exploiting the high yield strain of the POF compared to its glass counterpart which has been reported elsewhere [10]. Here, the segment of the fibre where stress is concentrated will result in the formation of a growing peak in the OTDR trace as the fibre experienced increasing strain.

2.2.2 Applications of OTDR-based POF sensors

Graded-index POF sensors in conjunction with the v–OTDR could be used for detection of cracks in tubular steel specimens encountered in the offshore oil platform. Small hairline cracks along weld lines are frequently the precursor to structural failure under the cyclic loading condition in which the structure is subjected to. Typically, offshore platforms are constructed using steel jacket structures. The steel jacket or 'substructure' is fabricated by joined together tubular steel members using welding and pin-jointed to the seabed. The steel jacket is a massive structure which can weigh up to 20,000 tonnes. In a complex structure such as that of an offshore platform, potential failure hot spots are many which may result from fatigue due to cyclic stresses. Cracking in the steel tubular members may occur due to the high stress intensity experienced at the intersections of brace and chord members. At certain critical sections, stress concentration would be as high as 20 times that of continuous non-welded sections. The welded joints and the abrupt change in geometry are likely locations where cracks would initiate and grow. An array of conventional non-destructive techniques such as ultrasonic scanning, X-radiography and acoustic emission exists for damage detection purposes but these are not suitable for real-time monitoring of large structures since these techniques, in general, require an experienced personnel to conduct the assessment in a localized manner. Hence a distributed sensing based on optical fibre sensor technology for crack detection offers a potential solution for health monitoring of large structures.

In a recent study conducted by Kuang et al., [17] scaled-size tubular steel specimens loaded under flexure was tested with the intention of using POF to monitor the crack growth in the specimen. A pre-crack was initially introduced and subsequently propagated under a three-point bend set-up. Graded-index POF was attached to the steel tube with the aim to detect and monitor the crack propagation as the applied load was increased in a quasi-static manner. In order to achieve this aim, a single POF was attached to the specimen in a serpentine manner such that as the crack propagated, different segments along the length of the fibre will be affected and will be reflected in the OTDR trace. The schematic of the specimen and photo of the set-up are shown in Figure 18. The POF (GigaPOF-50SR) tested had core and cladding diameters of 50 μm and 490 μm respectively. A 5 m long fibre was prepared in addition to a spacer fibre of 2.3 m. A pre-determined segment of the 5 m fibre was attached onto the polished section of the tube using a cynoacrylate-based adhesive. The fibre length from one position of crack to the adjacent one was pre-determined to be approximately 0.5 m (i.e. distance for each bent segment mid-point to mid-point), hence if the POF was able to detect the crack, a peak should appear at intervals of 0.5 m sequentially as the crack propagated.

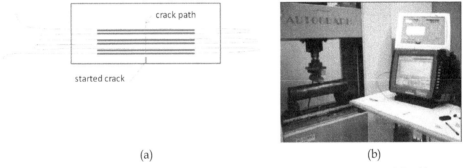

(a)   (b)

Fig. 18. (a) Schematic of the specimen, the bold lines indicating the segment of the fibre attached to the specimen (b) Photo showing the three-point bend set-up and the OTDR used in the study (after [17]).

The specimen was loaded transversely to propagate the pre-crack until the crack front reaches the vicinity of the first segment of the POF fibre. Any degradation or fracture of the plastic fibre was noted and compared to the OTDR plot. The OTDR was set to monitor the POF sensor continuously and updates the OTDR. Initially the measurement time of the OTDR was set to a minimum of 1 sec without signal averaging to assess whether this setting is sufficient to detect and locate the crack accurately to within 10 cm. Following the fracture of the first segment of the POF (i.e. closest to the crack front), the loading was momentarily stopped to allow the OTDR data to be saved for further inspection. The loading was continued subsequently until the crack propagates through the entire segment of the POF sensor. The locations of the fracture where they were expected to occur along the fibre were recorded and then compared to the OTDR results. The objective of the mechanical test was to observe the response of the POF and its potential as crack sensors on pre-cracked tubular steel tubes under transverse loading.

A close-up shot of the specimen following fracture of the POF sensors in all segments is shown in Figure 19. The photo reveals that the POF sensor underwent brittle fracture with little signs of necking and stress whitening. The sensing fibre was also noted to experience damage at approximately the same instance as when the crack front intersected the fibre. The damage of the sensing fibre was corresponded to a new peak in the OTDR trace. All the traces were consolidated and presented in Figure 20.

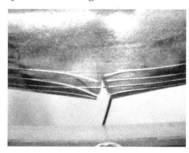

Fig. 19. Photo showing the fracture of the POF sensor along the crack path (after [17])

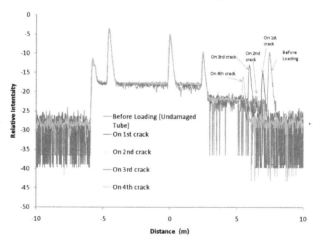

Fig. 20. Plot showing the consolidated OTDR traces highlighting the potential of the POF sensor in detecting and locating the position of the crack along the fibre (after [17]).

All the damage locations along the sensing fibre corresponding to the intersection of the crack front and the fibre were successfully located to within 20 mm by the v-OTDR except the second crack position. The authors argued that the POF segment at location 2 was not sufficiently sensitive i.e. the fibre at that particular location did not fracture when the crack intersected it. The sensitization of the fibre was clearly an important process to ensure that all cracks are detected. The adhesive used was also an important consideration as it is known that cynoacrylate-based adhesive embrittles the POF sensor and thereby improving its sensitivity to fracture.

Although the crack and location 2 was not detected in the particular test, it is evident that the technique shows excellent potential for detecting the location of cracks along the fibre and when configured in a serpentine shape, the sensor is capable of monitoring crack propagation accurately. From Table 1, it is clear that the v-OTDR possess the accuracy in determining the location of any crack along the POF given its high spatial resolution. The accuracy of locating the exact position of the crack along the fibre sensor relies on a proper calibration of the OTDR and determination of an accurate average refractive index of the graded-index POF. The results clearly showed that the calibration of the system was appropriate for the length of fibre used in the study.

Position	Expected (m)	Actual (m)
Location of 1st crack	6.964	6.95
Location of 2nd crack	6.464	-
Location of 3rd crack	5.964	5.95
Location of 4th crack	5.464	5.464

Table. 1. Accuracy of the crack locations detected by the OTDR versus actual positions (after [17]).

3. Conclusion

Plastic optical fibre sensors represent an emerging alternative for various applications in engineering. In this chapter, various examples of applications of POF sensors in civil engineering have been demonstrated e.g. for monitoring strain, deflection, liquid level, vibration, detection of cracks, and these have been applied to different materials including concrete, steel, geotextile and laminate composites. POF sensors offer many unique features that could be exploited to achieve a cost-effective sensing system. Potential users need to have an appreciation of these benefits and the capability of these sensors. Following an objective assessment of the capability of POF sensors together with a clear understanding of the sensing specification required in any particular project, POF sensors may proved to be the ideal candidate for the job.

4. Acknowledgment

Funding by A*STAR-MPA through the National University of Singapore CORE centre under joint Grants no. R-264-000-226-490 and no. R-264-000-226-305 is acknowledged.

5. References

[1] K. S. C. Kuang and W. J. Cantwell, "The use of conventional optical fibres and fibre Bragg gratings for damage detection in advanced composite structures—a review," Applied Mechanics Reviews, vol. 56, pp. 493–513, 2003.
[2] G. Zhou and L. M. Sim, "Damage detection and assessment in fibre-reinforced composite structures with embedded fibre optic sensors-review," Smart Materials and Structures, vol. 11, pp. 925–939, 2002.

[3] K. Peters "Polymer optical fiber sensors—a review" Smart Materials and Structures, vol. 20, Article ID 013002, 17 pages, 2011.

[4] K. S. C. Kuang, S. T. Quek, C. Y. Tan, and S. H. Chew," Plastic optical fiber sensors for measurement of large strain in geotextile materials," Advanced Materials Research, vol. 47–50, pp. 1233–1236, 2008.

[5] K. S. C. Kuang, S. T. Quek, M. Maalej, "Remote flood monitoring system based on plastic optical fibres and wireless motes" Sensors and Actuators A: Physical A, vol. 147 pp. 449–455, 2008.

[6] S. Kiesel, K. Peters, T. Hassan, and M. Kowalsky, "Calibration of a single-mode polymer optical fiber large-strain sensor," Measurement Science and Technology, vol. 20, Article ID 034016, 7 pages, 2009.

[7] S. Kiesel, K. Peters, T. Hassan, and M. Kowalsky, "Behaviour of intrinsic polymer optical fibre sensor for large-strain applications," Measurement Science and Technology, vol. 18, pp. 3144–3154, 2007

[8] K. Murphy, S. Zhang, and V. M. Karbhari, "Effect of concrete based alkaline solutions on short term response of composites," in Proceedings of the 44th International Society for the Advancement of Material and Process Engineering Symposium and Exhibition, vol. 44, pp. 2222–2230, 1999.

[9] K. S. C. Kuang, W. J. Cantwell, and P. J. Scully, "An evaluation of a novel plastic optical fibre sensor for axial strain and bend measurements," Measurement Science and Technology, vol. 13, pp. 1523–1534, 2002.

[10] P. Lenke, S. Liehr, K. Krebber, F. Weigand, and E. Thiele, "Distributed strain measurement with polymer optical fiber integrated in technical textiles using the optical time domain reflectometry technique," in Proceedings of the 16th International Conference on Plastic Optical Fibre, pp. 21–24, 2007

[11] K. S. C. Kuang and W. J. Cantwell, "The use of plastic optical fibre sensors for monitoring the dynamic response of fibre composite beams," Measurement Science and Technology, vol. 14, pp. 736–745, 2003.

[12] P. Cortes,W. J. Cantwell, K. S. C. Kuang, and S. T. Quek, "The morphing properties of a smart fibre metal laminate," Polymer Composites, vol. 29, pp. 1263–1268, 2008.

[13] K. S. C. Kuang, Akmaluddin, W. J. Cantwell, and C. Thomas, "Crack detection and vertical deflection monitoring in concrete beams using plastic optical fibre sensors," Measurement Science and Technology, vol. 14, pp. 205–216, 2003.

[14] K. S. C. Kuang, S. T. Quek, and M. Maalej, "Assessment of an extrinsic polymer-based optical fibre sensor for structural health monitoring," Measurement Science and Technology, vol. 15, pp. 2133–2141, 2004.

[15] K. S. C. Kuang, S. T. Quek, and W. J. Cantwell, "Morphing and control of a smart fibre metal laminate utilizing plastic optical fibre sensor and Ni-Ti sheet," in Proceedings of the 17th International Conference on Composite Materials, 2009.

[16] K. S. C. Kuang, C. Y. Tan, S. H. Chew and S. T. Quek, "Monitoring of large strains in submerged geotextile tubes using plastic optical fibre sensors" Sensors and Actuators A: Physical A, vol. 167, pp. 338–346, 2011.

[17] K. S. C. Kuang and C. G. Koh, "Structural damage detection of offshore platforms using plastic optical fibre sensors" Proceedings of the 7th Asian-Australasian Conference on Composite Materials (ACCM-7), 2010.

6

Mechanical Property and Strain Transferring Mechanism in Optical Fiber Sensors

Dongsheng Li, Liang Ren and Hongnan Li
Dalian University of Technology
China

1. Introduction

Optical fiber sensors (OFSs) have attracted considerable interests for their superior sensing abilities, especially due to their electromagnetic inference immunity and high sensitivity. However, OFS sensors based on bare fibers are fragile and easily damaged. For their safe use in engineering sensing, the glass core of optical fibers has to be coated with protective coatings, or to be bonded with adhesive materials, for instance, epoxy. Low elastic modulus of the soft coatings or bonding layer results in various shear stresses along the middle layer between the fiber core and the host structure to be measured. A portion of the host material strain is absorbed by the protective coatings when the strain transfers from the host material to the fiber core, and hence only a small fraction of structural strain is sensed. Consequently, the strains directly sensed by OFSs are different from actual structural strains. This is the main issue to be discussed in the chapter.

For the purpose to discuss the strain transferring problem, fundamental mechanical properties of OFSs are first presented. The chapter then develops an analytical model to derive the relationship between OFS sensed strains and actual structural strains. It is discovered that the strains, directly sensed by an OFS, have to be magnified by a factor (strain transfer rate) to reflect actual structural strains. This is, of course, of paramount interests for the application of optical fiber sensors. The factors that affect the efficiency of strain transfer on the optical fiber sensor are deduced and discussed in details based on the theoretical analysis. Critical adherence length is proposed as a measure to guarantee sufficient strain transfer rate based on the analytical model. Detailed analysis shows that the critical adherence length is the minimum length with which the fiber has to be tightly embedded to a structure for adequate sensing. Furthermore, theoretical results are verified through laboratory experimentation with the fiber Bragg grating sensors. Measured average strain transfer rates agree well with those calculated from the analytical model. Furthermore, the strain transfer rate of an OFS embedded in a multi-layered structure is consequently developed and the shear modulus of the host material on the influence of strain transmission is discussed.

For surface-bonded OFSs, optical fibers are adhered on the surfaces of structural members. Adhesive material still exists in-between and optical fibers are only partial buried. The strain readings of OFSs can be heavily affected by the thickness and mechanical properties of the adhesive and therefore it is important to determine how these factors affect the measurements. Numerical finite element simulations are reviewed for strain transferring mechanism of surface-bonded OFSs. In the end, the chapter summarizes current investigations on strain transferring mechanism of optical fiber sensors and practical packaging issues.

2. Fundamental mechanical properties of optical fibers

An optical fiber is a flexible, transparent fiber made of very pure glass (silica) not much wider than a human hair that acts as a waveguide, or "light pipe", to transmit light between the two ends of the fiber (Thyagarajan & Ghatak, 2007). Optical fiber typically consists of a transparent core surrounded by a transparent cladding material with a lower index of refraction.

Materials Properties (1)	Symbols (2)	Values (3)	Unit (4)
Young's modulus of silica fiber	E	7.3×10^{10}	Pa
Young's modulus of silicon coating	E	2.55×10^6	Pa
Poisson's ratio of silicon coating	v	0.17~0.48	-
Shear modulus of silicon coating	G	8.5×10^5	Pa
Strength of silica fiber	σ	$0.35{\sim}4.8 \times 10^{10}$	Pa
Breaking strain of silica fiber	ε	~10%	-
Density	ρ	2200	Kgm^{-3}
Coefficient of thermal expansion	f	0.54×10^{-6}	K^{-1}

Table 1. Typical mechanical properties of the optical fiber (Fernando et al., 2002; Measures, 2001)

Optical fibers are drawn from a molten glass perform. Although fluorozirconate, fluoroaluminate, and chalcogenide glasses as well as crystalline materials like sapphire, are used for optical fibers, silica is the most common material, especially for optical fiber sensing applications. Optical fibers made of fused silica are perfectly elastic until their breaking point. They are brittle and do not yield as do metals when overstressed. For the interests of sensing applications, key mechanical properties of silica fiber are listed in Table 1. Due to variation of manufacturing processes, mechanical properties of optical fibers may fluctuate around the nominate values in Table 1.

Although the strength of silica fiber is in excess of 4.8 Gpa (~7% strain), the practical stress limit is much lower due to the presence of many microscopic flaws. Proof tests are then carried out by manufactures to guarantee a safe operating stress and a predictable life. Typically, optical fibers are proof tested to 0.35 Gpa, corresponding to about 0.5% strain (Measures, 2001). For most sensor applications, the strength of silica fiber is much higher than that of structural members to be measured. Fatigue tests showed that optical fibers have a long working life when its operating stress is a small fraction of the proof stress. For most long term sensing application, OFSs can survive and are adequate.

Furthermore, fiber Bragg gratings (FBG) are the most common sensing elements among OFSs. Their strength is of considerable interests. Experiments showed that the formation of FBGs with UV light leads to a substantial weakening of the optical fiber (Feced et al., 1997). Controversy exists. Varelas et al. (1997) have demonstrated that producing FBGs with a continuous wave caused only minimal reduction in the strength of the optical fiber. Further investigations are needed to study the effects of fabrication of FBGs on the fiber strength.

3. Strain transferring mechanism for embedded OFSs

Conventionally, the values measured by optical fiber sensors were assumed to be actual structural strains (Baldwin et al., 2002). In fact, the strain measured by an OFS is different from actual host structure strain because of the difference between the optical fiber core modulus and the modulus of the fiber coating or the adhesive. Such strain difference depends on detailed packing measures of OFSs. Generally, there are three methods to integrate OFSs with host structures in terms of packing strategies: (1) direct integration, in which OFSs are directly embedded in or surfaced bonded to a host structure; (2) sensor-packaging integration, by which OFSs are first fixed in a small tube or bonded on the surface of a plate, and then the tube or the plate is anchored in the host structure; (3) clamping integration, in which an OFS is gripped at two ends by a bracket fixed on the host structure. No matter how the OFSs are packaged and integrated, the fiber core is brittle and has to be protected by adhesives or a coating layer in most sensing applications to avoid fiber breakage and to ascertain its long term stability. However, such a protection results in inconsistency between the fiber strain and the structural strain. Such discrepancies are neglected in many applications of OFSs by simply assuming that the fiber strain is consistent with the host structural strain (Friebele et al., 1999; Udd, 1995). Such assumption gives acceptable measurement results for the OFSs with long gauge length, in which the peak host strain can be fully transferred into the fiber strain, but cannot provide good measurement strains for short gauge OFSs, for instance, FBG sensors, in which the effect of the bonded fiber length on strain transfer between the fiber and host structure is significant (Galiotis et al., 1984). It is, thus, of primary importance to have a detailed knowledge of the relationship between the host structural strain and the fiber strain in order to correctly interpret structural strain from the strains sensed in the fiber.

Since the Young's modulus of the fiber is typically much larger than that of the adhesive or the coating as shown in Table 1, the axial elastic displacements of the fiber and the host material are different. In order to measure the strain of host materials accurately, there is a need to develop the relationship between the strain sensed by an OFS and the actual strain of the host material. Many researchers all over the world have made great efforts in this field and obtained notable achievements. Pak (1992) analyzed the strain transfer of a coated optical fiber embedded in a host composite, which is strained by a far-field longitudinal shear stress parallel to the optical fiber. In this case, the maximum shear transfer occurs when the shear modulus of the coating is the geometric mean of the shear moduli of the fiber and the host material. Ansari & Libo (1998) developed a simply model for evaluating the strain transfer percentage from the surrounding matrix to a length of optical fiber embedded in it under the assumption that the strain at the middle of the bonded fiber is equal to that of the host structure at the same position. Experiments were performed with a white-light fiber-optic interferometer to confirm the theory results. Li et al (2002) considered the coating as an ideal elasto-plastic material and deduced the strain transfer coefficients when the host material experiences tensile stress and compression stress respectively. Galiotis (1984) designed a polydiacetylene single fiber in an epoxy resin host material subjected to tensile strain along the fiber direction, and measured the strains at all points along the length of the fiber by the method of resonance Raman. They found that the axial strain in the fiber rises from a finite value at the end of the fiber to a fairly constant value at the central portion of the fiber, and that the axial strain at the midpoint of the fiber is lower than that applied to the host material, which can be approximately explained by the shear-lag model of Cox (1952).

This chapter mainly concerns the study of a single optical fiber embedded in a finite host structure, which is subjected to uniform axial stress. To derive the relationship between the strain measured by an optical fiber and the actual one experienced by the structure, a more realistic hypothesis is proposed in this chapter.

3.1 Theoretical approach

An optical fiber usually consists of three layers: fiber core, cladding and coating. The core diameter of a single mode fiber is usually 9μm, and the value for a multiple-mode fiber is 50μm (Fig.1). The coating, i.e., the outer layer of an optical fiber, often has an outer diameter of 250μm and an inner diameter of 125μm. The coating is usually made of plastics.

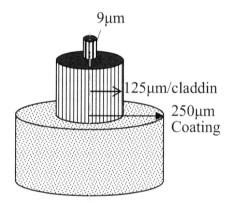

Fig. 1. Structure of a typical single mode fiber

We will deal with a concentric model of OFSs, which is similar to the fiber structure in Fig. 1. The concentric model is just a simplification of direct integration strategy. This model is suitable for two different cases. In the first case, a fiber with coating is directly embedded in the host structure. In the second case, the coating of a fiber is first stripped off, and then the bare fiber (including the core and cladding only) is attached to the inner surface of a steel tube by adhesives. In these two cases, there is an adhesive layer, often called middle layer, between the bare fiber and the host material. The three layers are concentric and the strain of the outer host material is transmitted to the inner fiber through the deformations of the middle layer.

For the concentric model, the strain transfer process from host material to the fiber core can be described as follows: as the host material generates strain under external loads, the deformation makes the shear strain appear in the interface between the host material and interlayer, and the shear strain is transferred to the interior of the interlayer to an extent which depends on the effective bonding of the interface. The interlayer generates axial strain at the same time. The axial strain in the interlayer is transferred to the fiber core through the shear strain in the interface between the interlayer and fiber core. The fiber core generates axial strain and senses the strain of the host material.

In the analysis of strain transfer to the optical fiber from the host material quantitatively, following assumptions are adopted for the concentric model.

Assumption 1) All the materials pertinent to the model remain elastic, and only the outer host material is subjected to axial stress and is uniformly strained, whereas the bare fiber and the middle layer do not directly bear any external loadings.

Assumption 2) Mechanical properties of the core and cladding of the fiber are the same. In reality, their properties are slightly different owing to their difference in some chemical components and the writing process of Bragg gratings. The core and cladding, hereinafter, are referred collectively as fiber for simplicity.

Assumption 3) There are no strain discontinuities across the interfaces, including the one between the host material and the middle layer and the one between middle layer and fiber interfaces, i.e. the bond between all the interfaces is perfect and no de-bonding exists.

In the concentric model, the Young's modulus of the single optical fiber is E_g with a radius of r_g embedded in a host material, separated by a middle layer of Young's modulus E_c and radius r_m and Poisson's ratio v, as illustrated in Fig.2. The host material, assumed to be infinite in all directions, is subjected to a uniform axial stress. L is the half gauge length of an optical fiber sensor, and $2L$ is the total length that the fiber is bonded to the host material through the middle layer. $\tau(x,r)$ is the shear stress in the middle layer a distance r above a given x coordinate along the center of the fiber. $\tau(x,r_g)$ is the shear stress at a given x coordinate along the fiber-middle layer interface. σ_g, σ_c and σ_m are the axial stress in the fiber, middle layer and the host material, respectively.

Based on the stress equilibrium for a small segment of the fiber and assumptions 1-3, the longitudinal stress along the fiber σ_g, is related to the interfacial shear stress at the fiber/middle layer interface through,

$$\frac{d\sigma_g}{dx} = -\frac{2\tau(x,r_g)}{r_g} \tag{1}$$

In reference to the free body diagram pertaining to the middle layer (Fig.2c), the equilibrium in the x direction results in the following relationship by balancing the shear stresses $\tau(x,r_g)$ and $\tau(x,r)$ with the axial stress σ_c,

$$\tau(x,r) = \frac{r_g}{r}\tau(x,r_g) - \frac{r^2 - r_g^2}{2r}\frac{d\sigma_c}{dx} \tag{2}$$

Substitution of (1) into (2) leads to,

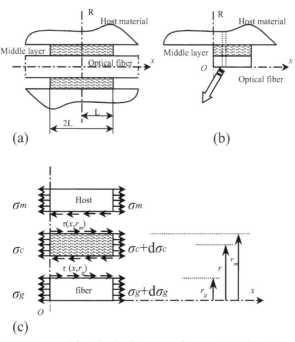

(c)

Fig. 2. Coordinate system and free body diagram of symmetrical section of optical fiber: (a) Optical fiber of gauge length 2L; (b) One quarter of the fiber; (c) Stress distribution of the fiber and the coating.

$$\tau(x,r) = -\frac{r_g^2}{2r}\frac{d\sigma_g}{dx} - \frac{r^2 - r_g^2}{2r}\frac{d\sigma_c}{dx} \tag{3}$$

Since the shear modulus of deformation predominates when it comes to the load transferring between the host material and the fiber, lateral motions perpendicular to the x coordinate are of secondary importance for the present study, and the Poisson effect can be neglected. Using the constitutive equation relating stress to strain, $\sigma_g = E_g \varepsilon_g$, where $\varepsilon_g = \frac{du}{dx}$ is the axial strain along the fiber and $u \equiv u(x)$ is the displacement, one obtains,

$$\begin{aligned}
\tau(x,r) &= -\frac{r_g^2}{2r}E_g\frac{d\varepsilon_g}{dx} - \frac{r^2 - r_g^2}{2r}E_c\frac{d\varepsilon_c}{dx} \\
&= -\frac{E_g r_g^2}{2r}\left(\frac{d\varepsilon_g}{dx} - \frac{r^2 - r_g^2}{r_g^2}\frac{E_c}{E_g}\frac{d\varepsilon_c}{dx}\right)
\end{aligned} \tag{4}$$

Since the fiber is strained together with the middle layer, the strain gradients are expected to be of the same order,

$$\frac{d\varepsilon_g}{dx} \cong \frac{d\varepsilon_c}{dx} \tag{5}$$

Thus, the important factor that determines $\tau(x,r)$ in (4) is the ratio of the stiffness between the fiber and the middle layer. The Young's modulus of the optical fiber coatings, in the case of a coated fiber, or the Young's modulus of typical structural epoxies, in the instance of a bare embedded fiber is, as usual, ten percent or less than that of the glass fiber, and r is not much larger than r_g (the middle layer is typically very thin to admit efficient strain transfer between the fiber and the host material), the second part of the right hand side of (4) is , therefore, insignificant compared to the first part,

$$\frac{r^2 - r_g^2}{r_g^2}\frac{E_c}{E_g}\frac{d\varepsilon_c}{dx} \cong o(\frac{d\varepsilon_g}{dx}) \tag{6}$$

Substituting (5) and (6) into (4) results in,

$$\tau(x,r) = -\frac{r_g^2}{2r}\frac{d\sigma_g}{dx} = -\frac{r_g^2}{2r}E_g\frac{d\varepsilon_g}{dx} \tag{7}$$

The shear stress term in (7) is determined by using Hooke's law as follows,

$$\tau(x,r) = G_c\gamma(x,r) = G_c(\frac{\partial u}{\partial r} + \frac{\partial w}{\partial x}) \cong G_c\frac{\partial u}{\partial r} \tag{8}$$

where $u \equiv u(x)$ and $w \equiv w(x)$ are the axial and radial displacements in the middle layer, respectively. Substituting (8) into (7) and integrating the resulting expression over r_g from the fiber and middle layer interface to the middle layer and host material interface radius r_m gives,

$$\int_{r_g}^{r_m} G_c\frac{\partial u}{\partial r}dr = \int_{r_g}^{r_m}\left[-\frac{r_g^2}{2r}E_g\frac{d\varepsilon_g}{dr}\right]dr \tag{9}$$

The integration result of (9) is,

$$u_m - u_g = -\frac{E_g}{2G_c}\frac{d\varepsilon_g}{dx}r_g^2\ln(\frac{r_m}{r_g}) = -\frac{1}{k^2}\frac{d\varepsilon_g}{dx} \tag{10}$$

where u_m and u_g respectively denote the axial displacement from the x coordinate origin in the host material and fiber. The strain lag parameter, k, containing both effects of the geometry and the relative stiffness on the system components, can be written as,

$$k^2 = \frac{2G_c}{r_g^2 E_g\ln(\frac{r_m}{r_g})} = \frac{1}{(1+\mu)\frac{E_g}{E_c}r_g^2\ln(\frac{r_m}{r_g})} \tag{11}$$

where, G_c, is the shear modulus of the middle layer. Differentiation of (10) with respect to x yields,

$$\frac{d^2\varepsilon_g(x)}{dx^2} - k^2\varepsilon_g(x) = -k^2\varepsilon_m \tag{12}$$

The general solution to (12) takes of the following form,

$$\varepsilon_g(x) = c_1 e^{kx} + c_2 e^{-kx} + \varepsilon_m \tag{13}$$

where c_1 and c_2 represent the integration constants determined by boundary conditions. Since the host material does not contact the fiber beyond the ends of the interface between the fiber and the middle layer, the fiber is assumed to be free from axial stress at both ends. This assumption means that the strain transferred to the fiber is equal to zero at both ends of the OFS, and is given by,

$$\varepsilon_g(L) = \varepsilon_g(-L) = 0 \tag{14}$$

The boundary conditions established by (14) are identical to the following equations,

$$\varepsilon_g(L) = 0, \ \dot{\sigma}_g(0) = 0, \dot{\varepsilon}_g(0) = 0 \tag{15}$$

The relationship (15) states that there is no shear stress in the fiber at the midpoint of the fiber (refer to (1)) due to the symmetric nature of the structure. The boundary conditions established in (14) and (15) will lead to the same solution as (13). Hence, by imposing these boundary conditions on (13), c_1 and c_2, are evaluated and obtained as

$$c_1 = c_2 = -\frac{\varepsilon_m}{2\cosh(kL)} \tag{16}$$

Thus, the final solution to (13), i.e. the strain distribution along fiber and its relationship to the strain of the host material at a given x coordinate is,

$$\varepsilon_g(x) = \varepsilon_m (1 - \frac{\cosh(kx)}{\cosh(kL)}) \tag{17}$$

Eq.(17) is the governing equation that describes the strain distribution along the fiber. The effects of the Young's moduli of the fiber and the middle layer, as well as the effects of the radii of the fiber and the middle layer, on the strain transfer, are all included in the strain lag parameter k defined in (11).

Fig.3 shows the strain transfer rate along an optical fiber. The mechanical properties of the optical fiber employed in the investigation, are given in Table 2. These properties are the same as those used by Ansari & Libo(1998) for comparison purposes. The strain difference between the fiber and the host material at a given x coordinate, is determined by the strain lag parameter k. Fig.3 also demonstrates that the strain sensed by the fiber at the midpoint is not equal to the strain in the host material in this instance.

3.2 Influence of the host material

Previous research on embedded OFSs have been carried out under the assumption that the strain distribution of the host material is not influenced by the embedding of the OFS and equals to the actual strain of the host material on the outer surface of the interlayer. According to Newton's third law, when strain in the host material causes deformations in

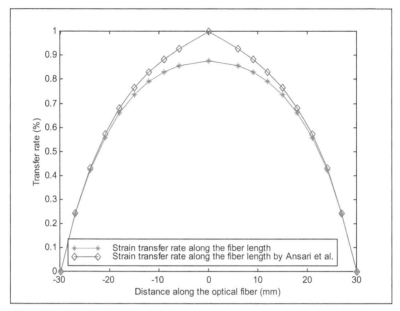

Fig. 3. Distribution of normal strain in fiber along the length

Materials Properties (1)	Symbols (2)	Values (3)	Unit (4)
Young's modulus of the fiber	E_g	7.2×10^{10}	Pa
Young's modulus of silicon coating	E_c	2.55×10^6	Pa
Poisson's ratio of silicon coating	μ	0.48	-
Shear modulus of silicon coating	G_c	8.5×10^5	Pa
Radius of outer boundary of silicon coating	r_m	102.5	μm
Radius of the fiber	r_g	62.5	μm

Table 2. Mechanical properties of the optical fiber

the interlayer, the interlayer reacts causing opposite deformations in the host material. Thereby, the strain level of some of the host material surrounding the embedded OFSs is lower than the actual strain of host material. The effects of host material properties in strain transfer are often ignored. However, many experiments and interrelated investigations reveal that this will result in large errors of strain transfer in the calculation especially when the stiffness of host materials is much lower than that of the fiber core.

In the field of short-fiber composites, Cox(1952) first developed the theoretical analysis of load transfer from the matrix to the fiber and this theory is referred to as the shear-lag analysis. The Young's modulus of matrix is considered as an important parameter. Subsequently, Monette et al. (1992), Rosen et al. (1965) and Tsai et al. (1990) deduced different improved strain transfer formulae using different assumptions based on the principle of shear-lag theory. In their formulae, the Young's modulus of matrix was introduced as a variable. All the results indicated that the strain transfer depends on the characteristics of the matrix. Monette et al (1992) investigated the strain transfer employing both analytical theory and computer simulation. The conclusions imply that the strain distribution in the fiber is expressed as a function of E_f/E_m , here E_f and E_m are the Young's modulus of fiber and matrix respectively; the stress transfer efficiency increases as the value of E_f/E_m increases. In the same paper (Monette et al. 1992), a general theory was outlined based on a shear-lag type approach for a three-phase composite material made of a single fiber plus an interphase region embedded in a soft matrix. The shear strain in the matrix exists from the interphase to a distance chosen far away in the matrix. The computer simulation results infer that the strain transfer is deeply influenced by the ratio, E_f/E_m . Grubb and Li (1994) used Raman spectroscopy to measure the axial stress distribution along a fiber during a quasi-static single fiber pull-out test. They obtained that the shear-lag constant can be expressed as a function of the matrix shear modulus and geometric terms. One of these terms is the effective interfacial radius, r_g , which is the radius for which the strain in the matrix equals the average matrix strain. Raman measurements indicate that r_g is small, only four times the fiber radius. This result is supported by polarizing microscopy. The model of Greszczuk (1969), which assumes a uniform shear within an effective interaction thickness, gives a similar result. Jiang et al. (1998) concluded that the conventional shear-lag theory has many drawbacks. Despite considering the strain transfer from matrix to fiber, the influence of the embedded fiber on the matrix is not taken into account by the shear-lag theory. They developed the theoretical formulae of fiber and matrix strain field distributions employed both shear-lag theory and elastic theory. The significance of the matrix is emphasized sufficiently in their deformation model. They not only considered the strain transfer from the matrix to the fiber, but also considered the influence of the fiber on the behavior of the matrix.

All the research findings indicate that the strain distribution of the host material surrounding the fiber is influenced by the fiber embedding; the character of the host material have important effects on the strain transferring from host material to fiber. Therefore, the mechanical properties of the host material must be taken into account when exploring the strain transferring from host material to fiber.

In our previous investigations(Li et al., 2006; LI et al., 2009), an improved strain transfer model is developed considering the influence of the host material. An additional assumption is that the normal stress and the shear stress are both existent in some range of the host materials and the influencing radius is equal to four times the external radius of the interlayer. The derivation is similar to the simple model and the final solution is in the same form as in Eq.(17) with the only exception that the shear lag parameter, k, takes a different expression as follows,

$$k^2 = \frac{2}{r_f^2 E_f \{\frac{1}{G_c}\ln(\frac{r_c}{r_f}) + \frac{1}{G_m}\left[\frac{r_m^2}{r_m^2 - r_c^2}\ln(\frac{r_m}{r_c}) - \frac{1}{2}\right]\}} \tag{18}$$

3.3 Average strain transfer rate

The strain transferred from the host material to an optical fiber varies with the different points along the gauge length of the fiber. Strain transfer rate (STR) $\alpha(x)$ can be defined as the ratio of the strain measured by an OFS sensor and the strain actually experienced by the host material at a given point of the fiber (a given x coordinate) as following,

$$\alpha(x) = \frac{\varepsilon_g(x)}{\varepsilon_m} = (1 - \frac{\cosh(kx)}{\cosh(kL)}) \qquad (19)$$

The maximum STR $\alpha_m(x)$ occurs at the midpoint of the fiber, i.e. at the point where x is equal to zero.

$$\alpha_m(0) = \frac{\varepsilon_g(0)}{\varepsilon_m} = (1 - \frac{1}{\cosh(kL)}) \qquad (20)$$

Generally, the strain measured by an OFS sensor, is the average strain over the gauge length of the fiber. Average STR (ASTR) can be expressed in the following form,

$$\overline{\alpha} = \frac{\overline{\varepsilon}_g(x)}{\varepsilon_m} = \frac{2\int_0^L \varepsilon_g(x)}{2L\varepsilon_m} = 1 - \frac{\sinh(kL)}{kL\cosh(kL)} \qquad (21)$$

The ASTR determined by (21) depends on the gauge length and the mechanical properties of the optical fiber, the middle layer and the host material. Therefore, it can be readily employed for correct interpretation of structural strains from the optical fiber measurement values. The results from the developed analytical model can be used as a complement to experiments especially where calibration tests are difficult to perform or where qualitative interpretation of a measurement system is required, for instance, in embedment applications of OFSs.

Fig.4 shows the variation of ASTRs $\overline{\alpha}$ over the fiber length and maximum STRs at the midpoint of the fiber in terms of the gauge length of the optical fiber, and also the experimental values by Ansari and Yuan (1998). It can be seen that the ASTRs and maximum STRs for the optical fiber in our study differs from those derived by Ansari and Yuan (1998), especially for the optical fiber with half-gauge length shorter than 60mm. When the fiber gauge length increases, the ASTRs and the maximum STRs determined by Ansari and Yuan (1998) are close to the values presented in this paper since sinh(kL) is nearly equal to cosh(kL) for a larger fiber half gauge length L. It should be emphasized that (21) is a more accurate solution for the strain transfer problem in the concentric cylinder model.

As indicated in Fig.4, the ASTR experimental values by Ansari and Yuan (1998) agree quite well with those evaluated by (21). Although the model has been verified for the simple case, it needs further experiments to validate its generalization. In any case, it facilitates a simple and direct qualitative interpretation of structural strains from measurement values made by OFSs.

Since the STRs at all the points within the fiber gauge length varies and a Bragg grating demands uniform axial deformations to avoid multiple-peak reflection and light spectrum expansion, it is required that an optical fiber be evenly stressed. However, only a short portion of FBG sensor is usually bonded, and this will lead to inadequate strain transfer, i.e. the strain sensed by a FBG sensor is only a fraction of the host material strain. Fig.5 shows

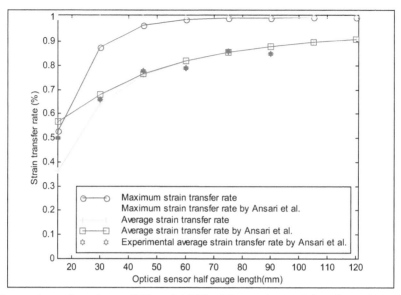

Fig. 4. Comparison of maximum STR and ASTR with experimental data for various gauge length sensors

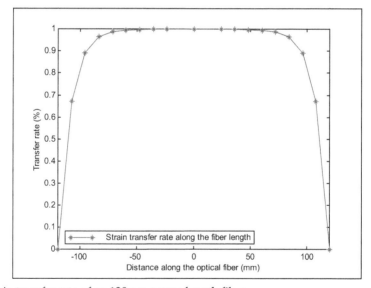

Fig. 5. Strain transfer rate of an 120mm gauge length fiber

the variation of ASTRs $\bar{\alpha}$ along the fiber with a gauge length 120mm. The figure indicates that the STRs near the center of the fiber approach unity. Therefore, an adequate fiber length has to be bonded to ensure the correct measurement of the host structural strain. Critical adherence length (CAL) can be defined as the minimum length to be bonded so that the STRs, at least over the middle half-length of the fiber, are larger than 0.9, i.e.

$$\alpha(l_c / 2) \geq 0.9 \tag{22}$$

Substituting (19) into (22) gives,

$$\frac{\varepsilon_g(l_c / 2)}{\varepsilon_m} = 1 - \frac{\cosh(kl_c / 2)}{\cosh(kl_c)} \geq 0.9 \tag{23}$$

The solution of (23) is,

$$\textbf{CAL}=l_c=9.24/k \tag{24}$$

where l_c is only computed over the half gauge length. Critical adherence length means that an OFS can be bonded over a longer portion, and the effective gauge length is located in the middle. For the case of the OFS shown in Table 2, its CAL is 99.78mm. This implies that the STRs along the middle half-length of the fiber are larger than 0.9 as long as the minimum bonded fiber length is beyond 99.78mm. **CAL** indicates that we can bond an FBG over a longer length, for instance 80mm, to locate the FBG just in the middle of the adhered length for efficient strain transferring.

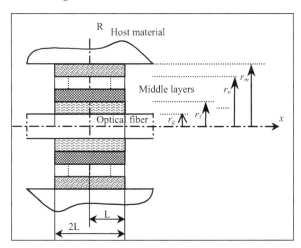

Fig. 6. Cross section of a multi-layer adhered fiber

3.4 Strain transferring for a multi-layered concentric model

In many cases, there are several middle layers between the fiber and host material, for instance, the fiber is first coated with an adhesive that solidifies quickly, and then bonded to the structure by an epoxy that solidifies slowly to ensure a uniform stress distribution. One typical case is to bond the fiber to a structure by directly applying adhesives on the fiber coating, and the coating and the adhesive form two separate layers between the fiber and host material. In some sensing applications, specialized coatings are required to enhance an optical fiber's measurement sensitivity and to accommodate the host structure. The OFSs maybe coated in this way with two different layers of coatings to employ their advantages of mechanical properties.

Fig.6 shows a multiple layer model available for the strain transfer analysis. Additionally, r_i is the outer radius of the ith layer (i =1~n), r_g is the outer radius of the fiber layer and r_m is the inner radius of the host materials (i.e. the outer radius of the (n+1)th layer). G_i is the shear modulus of the ith layer (i =2~n), and G_c is the shear modulus of the 1st layer.

Equation (9) can be rewritten as,

$$\int_{r_g}^{r_m}[\frac{dx}{dr}]dr = \int_{r_g}^{r_m}[-\frac{1}{G_c}\frac{r_g^2}{2r}E_g\frac{d\varepsilon_g}{dx}]dr \tag{25}$$

The resulting expression is integrated, over r_g, from the fiber and first layer interface, to the outmost layer and host material interface radius r_m through all the middle layers as follows,

$$(\int_{r_g}^{r_1}+\int_{r_1}^{r_2}+...+\int_{r_n}^{r_m}[\frac{dx}{dr}])dr = (\int_{r_g}^{r_1}+\int_{r_1}^{r_2}+...+\int_{r_n}^{r_m}[-\frac{1}{G_c}\frac{r_g^2}{2r}E_g\frac{d\varepsilon_g}{dx}]dr \tag{26}$$

The integration result of (25) is given by,

$$u_m - u_g = -\frac{1}{k_m^2}\frac{d\varepsilon_g}{dx} = -r_g^2\frac{E_g}{2}\{\sum_{i=2}^{n}\frac{1}{G_i}\ln(\frac{r_i}{r_{i-1}})+\frac{1}{G_c}\ln(\frac{r_1}{r_g})\}\frac{d\varepsilon_g}{dx} \tag{27}$$

where u_m and u_g denotes the axial displacement from the x coordinate origin in the host material and fiber, respectively.

$$k_m^2 = \frac{2}{r_g^2 E_g\{\sum_{i=2}^{n}\frac{1}{G_i}\ln(\frac{r_i}{r_{i-1}})+\frac{1}{G_c}\ln(\frac{r_1}{r_g})\}} \tag{28}$$

The strain lag parameter k_m, similar to the formerly discussed case, is determined by Young's moduli of the fiber and the middle layers, and the diameters of the fiber and the middle layers.

Thus, (19) and (28) can be directly used to compute the strain transfer rate for a fiber embedded in host material with multiple layers. Consequently, the critical adherence length and the average strain transfer rate within a gauge length for an optical fiber sensor can be computed by (19) and (28).

3.5 Test validation

The experiment is conducted on a universal material testing machine to study the strain transmission characteristics of an FBG sensor on different material sheets through a comparison of the strain values measured with the bare FBG and the electric strain gauge. A steel plate and a plexiglass plate are used in the experiments. Both plates are in rectangular shapes with uniform thickness. Two bare FBG sensors were bonded directly on one side of the plate with the two-component epoxy resin. One is located in the middle of the specimen and the other is located near the end of the specimen. Two electric strain gauges are fixed at the same positions on the other side of the specimen. The experimental specimen is shown in Figs. 7 and 8.

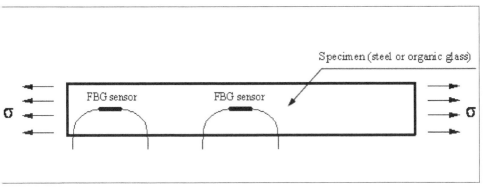

Fig. 7. The plan figure of the FBG sensors distribution.

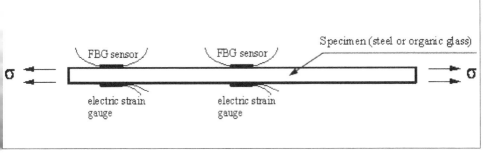

Fig. 8. The profile chart of sensor distribution.

Tensile experiments were carried out on the specimens. The load is controlled in the linear elastic range of the specimens. Therefore, each point of the specimens is under the same strain presumably from the point of view of elastic behaviour. The strain measured by the strain gauge is the actual strain of the plate. The ratio of the strain sensed by the FBG sensor to the strain measured by the strain gauge is considered to represent the strain transfer rate from the host material to the fiber core. As shown in Fig. 9, the steel plates is loaded step by step and continuously from 0 με to 250 με, then unloaded to 0 με. It is seen that the linearity of strain with the wavelength is very good, and that the coefficient of linear correlation is larger than 0.999. As shown in Fig. 10, the plexiglass sheet is loaded step by step and continuously from 0 μεto 500 με, and the coefficient of linear correlation is larger than 0.999 as well.

Based on the theory developed in this chapter, the average strain transfer ratios in steel and plexiglass as shown in Eq.(21) are 0.82330 and 0.7892 respectively. The ratio of average strain transfer with two different host materials is 0.9586. The experiment values can be calculated using test data. The experiments were repeated three times with three specimens and the data are listed in Table 3. All differences between the test and the theoretical values are less than 5%. Thus, the theoretical model is verified and the model is suitable for the analysis of fiber optic sensor strain transfer with different shear modulus of host materials.

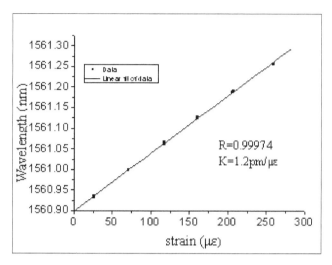

Fig. 9. the relationship between FBG wavelength and strain on the steel plate.

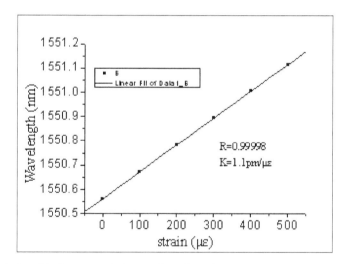

Fig. 10. the relationship between FBG wavelength and strain on the plexiglass plate.

Average strain transfer rate	First specimen	Second specimen	Third specimen
Experiment values (c_1)	0.9602	0.9596	0.9543
Theoretical values (c_2)	0.9586	0.9586	0.9586
Error ((c_1-c_2) / c_2)	0.2%	0.1%	0.5%

Table 3. Average strain transfer rates of bonded fiber Bragg grating sensors.

When the axis of the optical fiber has an angle with the direction of the applied principal strain/ stress, there would be a different strain/stress transfer rate existence because of the interlayer between the fiber optic sensor and the surrounding matrix. Li et al.(2007) discussed this aspect. Jiang & Peters (2008) developed a shear-lag model for unidirectional multi-layered structures whose constituents vary throughout the cross section through the extension of an existing optimal shear-lag model suitable for two-dimensional planar structures and discussed solution algorithms for a variety of boundary conditions.

4. Strain transferring mechanism for surface-bonded OFSs

Apart from embedded FBG sensors, bonded FBG sensors are used more frequently. Because of the asymmetry of the bonded sensors system, the theoretical analysis is difficult. The research methods concentrated on computer simulation with finite element model (FEM) mostly. Betz et al. (2003) studied strain transferring of patch bonded FBG sensors with finite element analysis and experiments. The fiber is first placed on a backing patch and this patch is simply glued to the surface of a structure. The thickness and the modulus of the backing material are varied in the model. When the structure is strained 0.3% parallel to the direction of the fiber axis, the strain level in the fiber core is found to vary between 0.26%~0.28%. It is concluded that the structural strain has not been completely transferred due to the fiber due to the presence of the backing patch, and that the degree to which the strain is transferred depends on the thickness and Young's modulus of the backing. Lin et al., (2005) investigated three packaging methods of FBG sensors and found that strain transmission rates decrease with the increase in thickness of the steel tube. Their experiments show, however, that the thickness and Young's modulus of the glues have little influence on the strain transmission.

Li et al. (2008) and Zhou et al. (2010) found also that the strain transmission loss is small when the substrate is thick and stiff as compared to the bonding layer and the FBG. However, it becomes large when the substrate is thin and made by low modulus materials. Wan et al. (2008) studied strain transfer with stiff adhesive case with a 3D-FEM for a surface-mounted strain sensor and verified it by experiments. They found that the bond length and the bottom thickness are dominant factors besides side width and top thickness. Jahani & Nobari (2008) demonstrated that both Young's and shear moduli of adhesive are frequency dependent in dynamic testing because adhesive materials show viscoelastic behaviour.

5. Conclusion

The strain transfer mechanism from host material to the fiber core can be described as follows: as the host material generates strain under the external load, the non-uniform deformation makes the shear strain appear in the interface between the host material and interlayer, and the shear strain is transferred to the interior of the interlayer to an extent which depends on the effective bonding of the interface. The interlayer generates axial strain at the same time. The axial strain in the interlayer is transferred to the fiber core through the shear strain in the interface between the interlayer and fiber core. The fiber core generates axial strain and senses the strain of the host material. In the course of strain transferring from host material to fiber core, a portion of strain is absorbed by the interlayer; the fiber core senses a partial strain of the host material only. Theoretical studies and experiments found that the structural strain may not be completely transferred due to the fiber core due to the presence of the coatings or packaging (middle layer). The strain transfer rate depends not only on geometry and mechanical properties of the middle layer or the material between the fiber and host surface, but on the Young's modulus of the host material. Naturally, packaging methods have significant influences on the strain transfer rate.

In practical applications, primitive protective coatings by fiber manufactures have to be stripped off for sufficient strain transfer rate, or at least its low Young's modulus has to be considered in experiments. Moreover, the aging problem of the adhesive should be considered since the service life of most structures to be measured is generally designed for 50 years or longer and adhesive must be adequate.

6. Acknowledgment

This research work was supported by the Science Fund for Creative Research Groups of the National Natural Science Foundation of China (Grant No. 51121005).

7. References

Ansari, F., & Libo, Y. (1998). Mechanics of Bond and Interface Shear Transfer in Optical Fiber Sensors. *Journal of Engineering Mechanics*,Vol.124, No.4,pp.385-394,ISSN 0733-9399.

Baldwin, C.;Salter, T.;Niemczuk, J.;Chen, P., & Kiddy, J. (2002). Structural monitoring of composite marine piles using multiplexed fiber Bragg grating sensors: In-field applications. *Smart systems for bridges, structures, and highways*,Vol.4696,pp.82,ISSN 0819444448.

Betz, D. C.;Thursby, G.;Culshaw, B., & Staszewski, W. J. (2003). Acousto-ultrasonic sensing using fiber Bragg gratings. *Smart Materials & Structures*,Vol.12, No.1,pp.122-128,ISSN 0964-1726.

Cox, H. (1952). The Elasticity and Strength of Paper and other Fibrous Materials. *British Journal of Applied Physics*,Vol.3, No.MAR,pp.72-79,ISSN 0022-3727.

Feced, R.;Roe-Edwards, M.;Kanellopoulos, S.;Taylor, N., & Handerek, V. (1997). Mechanical strength degradation of UV exposed optical fibres. *Electronics Letters*,Vol.33, No.2,pp.157-159,ISSN 0013-5194.

Fernando, G.;Webb, D., & Ferdinand, P. (2002). OPTICAL-FIBRE SENSORS. *MRS Bulletin*,Vol.27, No.5,pp.359-361,ISSN 0883-7694.

Friebele, E.;Askins, C.;Bosse, A.;Kersey, A.;Patrick, H.;Pogue, W., et al. (1999). Optical fiber sensors for spacecraft applications. *Smart Materials and Structures*,Vol.8, No.6,pp.813-838,ISSN 0964-1726.

Galiotis, C.;Young, R.;Yeung, P., & Batchelder, D. (1984). Study of model polydiacetylene/epoxy composites. Part 1- the axial strain in the fibre. *Journal of Materials Science*,Vol.19, No.11,pp.3640-3648,ISSN 0022-2461.

Greszczuk, L. Theoretical Studies of the Mechanics of the Fiber-Matrix Interface in Composites.ISSN 0091-3286.

Grubb, T., & Li, T. (1994). Single-fiber polymer composites: Part I Interfacial shear strength and stress distribution in the pull-out test. *J. Mater. Sci,*Vol.29,pp.189˜C202,ISSN.

Jahani, K., & Nobari, A. (2008). Identification of dynamic (Youngs and shear) moduli of a structural adhesive using modal based direct model updating method. *Experimental Mechanics,*Vol.48, No.5,pp.599-611,ISSN 0014-4851.

Jiang, G., & Peters, K. (2008). A shear-lag model for three-dimensional, unidirectional multilayered structures. *International Journal of Solids and Structures,*Vol.45, No.14-15,pp.4049-4067,ISSN 0020-7683.

Jiang, Z.;Lian, J.;Yang, D., & Dong, S. (1998). An analytical study of the influence of thermal residual stresses on the elastic and yield behaviors of short fiber-reinforced metal matrix composites. *Materials Science and Engineering: A,*Vol.248, No.1-2,pp.256-275,ISSN 0921-5093.

Li, D. S.;Li Sr, H.;Ren, L., & Song, G. (2006). Strain transferring analysis of fiber Bragg grating sensors. *Optical Engineering,*Vol.45, No.2,pp.4402,ISSN 0091-3286.

LI, H. N.;Zhou, G. D.;Ren, L., & LI, D. S. (2009). Strain Transfer Coefficient Analyses for Embedded Fiber Bragg Grating Sensors in Different Host Materials. *Journal of Engineering Mechanics,*Vol.135, No.12,pp.1343-1353,ISSN 0733-9399.

Li, H. N.;Zhou, G. D.;Ren, L., & Li, D. S. (2007). Strain transfer analysis of embedded fiber Bragg grating sensor under nonaxial stress. *Optical Engineering,* Vol.46 ,pp.054402,ISSN 0091-3286.

Li, Q.;Li, G.;Wang, G.;Ansari, F., & Liu, Q. (2002). Elasto-Plastic Bonding of Embedded Optical Fiber Sensors in Concrete. *Journal of Engineering Mechanics,*Vol.128, No.4,pp.471-478,ISSN 0733-9399.

Li, W. Y.;Cheng, C. C., & Lo, Y. L. (2008). Investigation of strain transmission of surface-bonded FBGs used as strain sensors. *Sensors & Actuators: A. Physical,*ISSN 0091-3286.

Lin, Y. B.;Chang, K. C.;Chern, J. C., & Wang, L. A. (2005). Packaging methods of fiber-Bragg grating sensors in civil structure applications. *Sensors Journal, IEEE,*Vol.5, No.3,pp.419-424,ISSN 1530-437X.

Measures, R. M. (2001). *Structural monitoring with fiber optic technology* ISBN 0124874304: Academic.

Monette, L.;Anderson, M.;Ling, S., & Grest, G. (1992). Effect of modulus and cohesive energy on critical fibre length in fibre-reinforced composites. *Journal of Materials Science,*Vol.27, No.16,pp.4393-4405,ISSN 0022-2461.

Pak, Y. (1992). Longitudinal Shear Transfer in Fiber Optic Sensors. *Smart Materials and Structures(UK),* No.1,pp.57-62,ISSN 0964-1726.

Rosen, B. W. (1965). Mechanics of composite strengthening. *Fiber composite materials,*Vol.2,pp.37-75,ISSN 0091-3286.

Thyagarajan, K., & Ghatak, A. K. (2007). *Fiber optic essentials* ISBN 0470097426: Wiley-IEEE Press.

Tsai, H. C.;Arocho, A. M., & Gause, L. W. (1990). Prediction of fiber-matrix interphase properties and their influence on interface stress, displacement and fracture toughness of composite material. *Materials Science and Engineering: A,*Vol.126, No.1-2,pp.295-304,ISSN 0921-5093.

Udd, E. (1995). *Fiber optic smart structures* ISBN 0471554480: Wiley-Interscience.

Varelas, D.;Limberger, H., & Salathe, R. (1997). Enhanced mechanical performance of singlemode optical fibres irradiated by a CW UV laser. *Electronics Letters*,Vol.33, No.8,pp.704-705,ISSN 0013-5194.

Wan, K. T.;Leung, C. K. Y., & Olson, N. G. (2008). Investigation of the strain transfer for surface-attached optical fiber strain sensors. *Smart Material Structures*,Vol.17, No.3,pp.5037,ISSN 0964-1726.

Zhou, J.;Zhou, Z., & Zhang, D. (2010). Study on Strain Transfer Characteristics of Fiber Bragg Grating Sensors. *Journal of Intelligent Material Systems and Structures*,ISSN 1045-389X.

7

Plastic Optical Fiber pH Sensor Using a Sol-Gel Sensing Matrix

Luigi Rovati[1], Paola Fabbri[2], Luca Ferrari[1] and Francesco Pilati[2]
[1]Department of Information Engineering,
[2]Department of Materials and Environmental Engineering,
University of Modena and Reggio Emilia
Italy

1. Introduction

Because it is the most ubiquitous species encountered in chemical reactions, hydrogen ion occupies a very special place in chemistry and biology, most of the chemical and biological processes being dependent on its activity. From the analytical point of view, the abundance of hydrogen ions is quantified in terms of pH, the negative logarithm of its activity. Its importance is evident by considering that, if the pH of the human blood changes as little as 0.03 pH units or less, the functioning of the body will be greatly impaired; also, brain pH decreases from normal pH of 7.4 to a pH of 6.75 during the brain insult and a continuous monitoring system would be beneficial in the treatment of comatose neurosurgical patients and those who have suffered traumatic brain injury, ischemic brain insult and so forth (Zauner, 1995). Furthermore, the kind of animals and plants living in lakes, rivers and oceans depends on pH values, as well as pH of soil affects the livability of plants. For this reason, the use of pH sensors is widely diffused in various fields to monitor chemical and biological processes and it is finding an increasing number of applications in medicine, biomedicine, industry and environmental monitoring.

The earliest methods of pH measurement fall roughly into four categories: indicator reagents, pH test strips, amperometric or potentiometric devices. Since a long time, the glass electrode pH sensor has been the most popular due to its ideal Nernstian response, independent of redox interferences, and because of its high selectivity for hydrogen ions in a solution, short balancing time of electrical potential, reliability and high reproducibility, and long lifetime. For these reasons, it is still considered to be the standard measuring method for the pH measurements.

However, glass electrodes have several drawbacks for many applications. Firstly, they exhibit a sluggish response and are difficult to miniaturize. Moreover they can generate electromagnetic interferences with other devices and cannot be used in food or in vivo applications due to their brittle nature. For these reasons, there is increasing need for alternative pH sensors and, in this respect, optical methods offer many advantages over the common glass electrode or other electrochemical devices. Among these, optical fiber colorimetric pH sensors monitoring the change in the optical response of proton-sensitive indicators are the most popular systems. Other pH-sensing devices, based on different pH-related properties, have also been proposed; they exploit mass-changing in pH-responsive hydrogel (Ruan, 2003), pH-sensitive changes of refractive-index in polymers (Zamarreno, 2010), pH sensing based on the interferometric response of nanofilms built from layer-by-layer electrostatic self-assembly of polyelectrolytes (Goicoechea, 2009), pH sensors based on conducting polymers (including polypirrole, polyaniline, etc.) (Talaie, 2000), ultrasensitive microcantilever structures (Bashir, 2002; Fritz, 2000), to mention only someone. However, these devices suffer from various drawbacks such as instability or drift or their use is not straightforward or are too expensive to be mass-produced at industrial level.

For a more detailed discussion about pH sensors, one may refer to several reviews appeared in the last decade (Lin, 2000; Wolfbeis, 2004; Yuqing, 2005; Korostynska, 2007; Baldini, 2008; Wolfbeis, 2008; McDonagh, 2008).

As said above, optical fiber methods are the most interesting alternative to the glass electrode for pH as they enable optical spectroscopy to be performed on sites inaccessible to conventional spectroscopy and over long distances. This is particularly interesting in medicine where the high degree of miniaturization of optical fibers, their considerable geometrical versatility, and their extreme handiness enable unique performances in invasive applications. With respect to glass electrode pH sensors, optical fiber methods take also advantage of the immunity to electromagnetic interferences, and from the absence of electrical contacts that make them useful also in the case of non-invasive applications. For these reasons, optical fiber sensors are in continuous development and offer to the physicians efficient tools for prompt and reliable diagnosis, in particular when an in-vivo continuous monitoring has to be carried out (Baldini, 2008).

The basic concept of the optical fiber methods of pH measurements relies on the fact that the incident beam of light is passed through a light guide to the active end of the optical fiber where it interacts with a chemical indicator, which alters the beam's intensity, usually by absorption or fluorescence, of the modified optical signal which is guided back to the detector. In this way, remote sensing can be achieved since the optical signal can be carried over long distances.

The indicator is usually an organic dye able to change its color over a given pH range (typically 3-4 pH units centered at the pKa of a given indicator) and is generally confined to the surface of the optical fiber or immobilized in an adjacent space according to various possible configurations of the sensing element (Lin, 1997; McDonagh, 2008).

In order to record reliable pH values, the hydrogen ion concentration must equilibrate between the bulk of the sample and the space where the indicator is confined, and the sensor response in terms of precision, sensitivity, stability and response-time will therefore depend on the conditions existing within this space.

The most used method exploited to prepare an optical fiber responsive to pH changes is the immobilization of pH indicators in/on a suitable material. This is a key step that will largely influence the characteristics of the pH sensor.

Photopolymerization (Barnard, 1991; Bronk, 1994; Ferguson, 1997; Healey, 1997; Song, 1997) and sol-gel chemistry (Lin, 1997 and 2000; McDonagh, 2008) have been the mainly exploited reactions, and adsorption, covalent binding and entrapment are the most widely used approaches to immobilize indicator dyes.

In the adsorption method, a pH indicator is adsorbed, physically or chemically, on a solid support attached to the optical fiber; it is relatively simple, but the amount of the adsorbed indicator may be small, resulting in a weak signal, and the indicator may leach out, resulting in a pH response that changes with time. Commercial ion-exchange resins, sulphonated polystyrene (Igarashi, 1994) and polyelectrolyte containing silica (Shi, 1997) have been proposed, among the other, for electrostatic immobilization, while hydrophobic organic products (polymers or organically modified alkoxysilanes) have been used to improve hydrophobic chemical adsorption of indicators (Korostynska, 2007; Kowada, 2005; Wu, 2006).

Adsorption of the indicator may be slow in non-porous monolithic solids and in solids with random microporosity, and the development of mesoporosity, having large and controlled open pores (Cagnol, 2004; Grosso, 2004), may result in a shorter response time of the pH sensor (enhanced diffusion and accessibility for protons and therefore faster equilibration) and a larger amount of pH indicator to be immobilized. However indicator leaching may also become a more relevant problem.

The covalent method can overcome leaching problems as the pH indicator is strongly bound to the solid substrate (Baldini, 2008; Barnard, 1991; Bronk, 1994; Ferguson, 1997; Healey, 1997; Song, 1997). However, the immobilization by covalent attachment requires a previous surface modification of the optical fibers and the control of the surface reaction of the modified substrate with a suitable indicator. The resulting procedure is often complicated and time consuming and may lead to loss of indicator sensitivity or result in poor fluorescent properties (Bacci, 1991; Lobnik, 1998). In general, this leads to a costly technology unsuitable for the preparation of mass-produced disposable devices.

In the entrapment method, the pH indicator is entrapped in a substrate, often starting from a polymer solution or a polymerizable liquid. The method is relatively easy, and a suitable choice of the indicator/polymer pair can strongly reduce or completely overcome the problem of indicator leaching. In addition, polymers present the advantage to be less brittle than inorganic glasses, and a suitable design of the molecular architecture can lead to a good balance of hydrophobicity and water swelling. This could allow combining a faster penetration of water and ions with a minimized indicator leaching. Various polymers have been used for entrapment of pH indicators by exploiting either their thermodynamic affinity or by covalent binding of pH indicators (Korostynska, 2007; McDonagh, 2008).

Combination of adsorption and entrapment methods is expected to occur in polymer-containing organic-inorganic hybrid materials, where highly interpenetrated organic and inorganic nanosized domains are formed during the sol-gel process. This could result in improved sensors, when, in addition to the control of water swelling by a suitable polymer choice, the presence of an extended interfacial surface between organic and inorganic domains, can favor the physical adsorption of the indicator. This can allow to optimize the mobility of the hydrogen ions, from the sample into the sensing element, resulting in a reduced response time, and to minimize indicator leaching.

Optical fiber pH sensors can also be prepared in the absence of indicators by exploiting the change of color as a function of pH of some conductive polymers such as polyanilines (Pringsheim, 1997; Sotomayor, 1997; Grummt, 1997) and polypyrrole (de Marcos, 1996). Even though they present some intrinsic advantages, such as simple fabrication and measurements in the near-IR region, they suffer from shortcomings like long response time (slow diffusion of water and hydrogen ions into the glassy polymers), interference from other ions, oxidation by oxidizing agents and the need for pre/reconditioning before each measurement.

In the last two decades sol-gel chemistry has become one of the most widely used method for the development of optical fiber pH sensors due to its versatility and to the attractive optical properties of inorganic sol-gel materials (Lin, 1997, 2000; Wolfbeis, 2004, 2008; McDonagh, 2008). Indeed, either inorganic glasses or organic-inorganic hybrid materials can be easily obtained under mild conditions (T < 100 °C) starting from a low-viscosity solution containing metal alkoxides alone or mixed with organically modified alkoxysilanes or with suitable polymers. Sol-gel process key parameters like reactant formulation, catalyst, water content and reaction temperature, can be exploited to produce a large variety of materials whose properties can be finely adjusted to tailor the physicochemical properties of the final solid in order to optimize sensor performances (Wen, 1996; Cagnol, 2004; Grosso, 2004; Estella, 2007).

Various sol-gel material configurations, like monoliths, thin films, capillaries and droplets including miniaturized configurations, have been proposed to generate the sensing element at the optical fiber tip (Lin, 1997; McDonagh, 2008).

Alkoxysilanes are most frequently used as precursors of inorganic phases, and the resulting inorganic silica glasses are porous matrices, that can be characterized by tailored porosity (Wen, 1996; Cagnol, 2004; Grosso, 2004; Estella, 2007). Compared with pure organic polymers, sol-gel inorganic glasses can offer some advantages: higher stability, optical transparency down to 250 nm, feasibility of direct coating on glass and silica fibers. However, they also may suffer from some shortcomings, and in particular, slow response to pH changes (typically several tens of seconds) (Lin, 1997; Ben-David, 1997; Seki, 2007), indicator leaching, difficult adhesion to plastic optical fibers.

Sol-gel materials derived from organically modified alkoxysilane (Ormosil) can be prepared under the same reaction conditions used for the preparation of all inorganic sol-gel glasses. The resulting material can have a better surface interaction with organic indicators reducing the shortcomings deriving from leaching (Kowada, 2005; Wu, 2006).

The inclusion of suitable organic polymers in the initial solution with alkoxysilanes leads to the formation of silica-based hybrid materials consisting of highly interpenetrated organic and inorganic domains. A suitable choice of the organic polymers will result in the formation of organic domains that can allow to immobilize higher amounts of indicator and to optimize indicator performances (Cjlakovic, 2002; Aubonnet, 2003) versus water adsorption, while preserving good transparency. As it will be discussed in the following, this can result in fast and reliable cheap sensors.

A lot of progress, improvement and innovation have been made in the field of optical fiber sensors in the last two decades, however notwithstanding the high number of optical fiber chemical and biochemical sensors described in the literature, only a few have reached the market, and the goal of mass production and commercialization of optical fiber pH sensors has not really been accomplished, and research development and application of optical fiber pH sensors is still very challenging and demanding interdisciplinary field. To reach this goal issues such as low cost, user-friendliness, sensitivity, robustness, long-term stability and shelf lifetime of the sensors have to be addressed adequately.

Within this context, this paper aims at illustrating a facile method to develop plastic optical fiber pH sensors with a tip-based sensing element prepared by a sol-gel process and consisting of phenol red indicator entrapped in a PEO-silica organic-inorganic hybrid material.

2. Principle, theory, design of the POE pH sensor

Current trends in optical sensors, such as miniaturization, flexibility and enhanced sensitivity, are indicating a new chemical route for the development of advanced multifunctional materials for optical applications. Those chemical technologies, which can be more easily customized and allows the inclusion of multiple functionalities within a unique preparation step, are bound to be progressively more and more applied to the preparation of optical materials.

In this perspective, the sol-gel technology certainly represents one of the most promising chemical strategies, thanks to numerous advantages mainly related to simplicity and mild operative conditions. It enables creating a glass-like porous structure at room temperature by a two-step acid or base catalyzed reaction involving hydrolysis and condensation, starting with metal alkoxides $M(OR)_4$, which transforms into a rigid three-dimensional metal-oxide network (Brinker, 1990). The sol–gel process has been proved to be flexible enough for an efficient incorporation of organic polymer chains that can behave as flexible links between the metal-oxide domains in the inorganic network, in particular when they are bearing reactive groups that can be involved in the hydrolysis–condensation reactions. The resulting materials are known as organic–inorganic hybrids (Schmidt, 2000), also commonly designated as ceramers due to the combination of the properties of ceramics (high modulus, thermal stability and low coefficient of thermal expansion) with those of organic polymers (high ductility, molecular flexibility and low temperature processing). These materials are often also known as phase-interconnected nanocomposites because of the high level of interconnection between the two phases with domain phase sizes approaching the nanometer scale. Ceramers have a huge potential for application in a variety of advanced technologies (Eckert, 2001; Sanchez, 2011; Kickelkick, 2006), both as structural materials and functional materials, such as catalyst supports, protective coatings (Messori, 2003, 2004a); Toselli, 2007; Fabbri, 2008), sensors (Rovati, 2011; Fabbri, 2011), and active glasses.

Optical fiber sensors are traditionally obtain by fully-inorganic sol-gel process that allows the creation of Si-O-Si linkages between the silica core of the optical fiber and the silica porous matrix deriving from the jellification of the sensitive dye-doped colloidal suspension (Cao, 2005). However, this approach cannot be easily applied in the case of plastic optical fibers, due to the ineffective interaction between the organic PMMA optical fiber core (Lin, 2000).

The approach proposed in this work consists in the fabrication of a pH sensor based on an organic-inorganic hybrid matrix obtained by a sol-gel process, doped with a pH sensitive indicator, to be applied at the tip of plastic optical fibers. Inside the sensitive element, the organic part of the hybrid glass, polyethylene oxide (PEO), plays a multiple role: (i) it allows good adhesion between the plastic optical fiber and the whole sensitive element; (ii) its weak hydrophilicity permits to tune the kinetic of response of the sensor by influencing the diffusion rate of the analyte inside the porous matrix and its interaction with the indicator; (iii) its nature of organic compound allows better physical and chemical interactions with the organic pH indicator dispersed in the hybrid matrix, thus reducing problems of leaching and enhancing the response rate of the sensor.

2.1 Polymer-silica sol-gel hybrids for pH indicators entrapment

Several studies have been performed on the preparation of pH sensors obtained by immobilization of a pH sensitive indicator onto the tip or onto the sides of traditional silica core optical fibers by means of the sol–gel technology (MacGraith, 1991; Gupta, 1998; Alvarado-Mendez, 2005; Rayss, 2002; Miled, 2002; Dong, 2008; Lee, 2001), thanks to the outstanding flexibility of this chemical technique. For this application, silica glass is usually prepared by sol–gel via hydrolysis and condensation reactions of the precursor tetraethylorthosilicate (TEOS), in the presence of water (to promote hydrolysis), ethanol (to enhance the miscibility of water and TEOS), and an acid or base catalyst. The resulting silica-based glass can be attached to the silica core of traditional optical fibers by uncladding of a portion of the optical fiber and activation of the silica core by immersion in nitric acid in order to form silanol groups onto the optical fiber surface, followed by deposition of the silica sol containing the pH indicator on the naked part of the optical fiber (Sharma, 2004). This procedure allows the creation of Si–O–Si linkages between the silica core of the optical fiber and the silica porous matrix deriving from gelification of the sensitive dye-doped colloidal suspension, thus achieving a permanent bonding between the sensitive element (sol–gel glass doped with indicator) and the optical fiber itself which will transmit the optical signal.

The nature of organic compound of the pH sensitive indicators can cause a scarce affinity with the inorganic silica matrix (Wu, 2006). This induces leaching out of the sensitive species from the support, and the only effective way to overcome this problem is to create covalent bonding or stabilizing interactions between the pH indicator and the embedding matrix; nevertheless, this approach demonstrated to strongly reduce the indicator's sensitivity because of the reduction of its mobility within the solid matrix, and to cause slow response.

If the same approach described above for the preparation of pH optical sensors applied to silica-core optical fibers would be applied to the new generation of plastic optical fibers, an ineffective interaction between the organic PMMA core and the inorganic silica sensitive element would be found, resulting in an unstable device with physical discontinuity between the optical fiber and the sensitive head applied onto its tip.

2.2 Plastic optical fibers and sensing probe

The developed sensor is based on a low-cost plastic optical fiber terminated with a sensing element. The sensing matrix takes advantage of the use of PEO/silica organic-inorganic hybrids obtained from sol-gel technique, and in particular of the weak hydrophilicity of PEO chains crosslinked by sol-gel reactions with TEOS. In the presence of aqueous solutions, the PEO domains can swell allowing a faster ion exchange with the surrounding solution. The extent of swelling depends on the PEO molecular weight and, in addition to the beneficial effect on ion diffusion rate, it leads to changes of the volume of the sensing element that can causes problems to the extraction of the pH information. A good balance of ion diffusion rate, adhesion to PMMA fibers and volume change was obtained using a PEO having a molecular weight 8000. A schematic diagram of the probe configuration is reported in Fig. 1. Basic working principle is based on a classic colorimetric approach. The white interrogation light $I_i(\lambda)$ is scattered and absorbed by the sensing matrix that include a pH sensitive indicator (indicator). The change in color of this indicator induces a change in the optical spectrum of the outgoing light $I_o(\lambda, pH)$ collected by the fiber. Analyzing the spectrum of this optical signal the pH of the solution can be recovered.

Absorption operated by the indicator, which depends on the pH value of the solution, induces a change in the optical spectrum of the outgoing light collected by the fiber.

As the pH varies, the relative fractions concentrations of the dissociated [A-] and protonated [HA+] forms of the indicator are changed according to the Henderson-Hasselbalch equation:

$$[HA^+] = \frac{C}{1 + 10^{pH - pK_a}}$$

$$[A^-] = \frac{C}{1 + 10^{pK_a - pH}}$$

(1)

where Ka is the acid dissociation constant, pKa is $-\log$Ka, and C is the concentration of the indicator. According to Fig. 1, interrogation light $I_i(\lambda)$ is absorbed and scattered by the sensing matrix generating the outgoing light $I_o(\lambda, pH)$. According to the modified Beer-Lambert law, the spectrum of the outgoing light $I_o(\lambda, pH)$ can be estimated from the spectrum of the entering light $I_i(\lambda)$ (Splinter, 2006):

$$OD(\lambda, pH) = Log\frac{I_i(\lambda)}{I_0(\lambda, pH)} = \alpha_A(\lambda) \cdot d(\lambda, pH) \cdot [A^-] + \alpha_{HA}(\lambda) \cdot d(\lambda, pH) \cdot [HA^+] + G$$

(2)

$$OD(\lambda, pH) = C \cdot d(\lambda, pH) \cdot \left\{ \frac{\alpha_A(\lambda)}{1 + 10^{pK_a - pH}} + \frac{\alpha_{HA}(\lambda)}{1 + 10^{pH - pK_a}} \right\} + G$$

where OD(λ, pH) is the optical density, $\alpha_A(\lambda)$ and $\alpha_{HA}(\lambda)$ the specific extinction coefficients of the two forms, d(λ, pH) the mean optical path of the photons through the probe, and G a parameter related to the "geometry" of the probe. Hence, we consider the relative variations of the outgoing light as the sensor response:

$$\Re(\lambda, pH) = \frac{I_0(\lambda, pH_{ref}) - I_0(\lambda, pH)}{I_0(\lambda, pH)} = -\frac{\Delta I_0(\lambda, pH)}{I_0(\lambda, pH)} = \frac{\Delta OD(\lambda, pH)}{Log(e)}$$

$$\Re(\lambda, pH) = \frac{C \cdot d(\lambda, pH)}{Log(e)} \cdot \left\{ \frac{\alpha_A(\lambda)}{1 + 10^{pK_a - pH}} + \frac{\alpha_{HA}(\lambda)}{1 + 10^{pH - pK_a}} \right\} - Bl(\lambda, pH_{ref})$$

(3)

where baseline Bl is

$$Bl(\lambda, pH) = \frac{C \cdot d(\lambda, pH_{ref})}{Log(e)} \cdot \left\{ \frac{\alpha_A(\lambda)}{1 + 10^{pK_a - pH_{ref}}} + \frac{\alpha_{HA}(\lambda)}{1 + 10^{pH_{ref} - pK_a}} \right\}$$

(4)

and pH$_{ref}$ is the reference value of pH that annuls response $\Re(\lambda, pH)$.

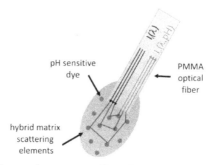

Fig. 1. Drawing of the probe configuration: $I_i(\lambda)$ is the interrogation light absorbed and scattered by the sensing hybrid matrix; $I_0(\lambda, pH)$ is the outgoing light containing the information of interest.

3. Prototype sensors

The developed sensor is based on a low-cost plastic (PMMA) optical fiber (Fort Fibre Ottiche, Italy) with core diameter 1 mm and numerical aperture 0.22. The total length of the taken optical fiber is 35 cm. The pH sensitive indicator (phenol red) was immobilized on the tip of the optical fiber taking advantage of the use of PEO to make hybrid gels. In the next sections the details about the prototype preparation are provided.

3.1 Chemicals

α,ω-Hydroxy-terminated poly(ethylene oxide) (PEO; purchased by Fluka, Milan, Italy) having a number-average molecular weight of 8000 g/mol, 3-isocyanatopropyltriethoxysilane (ICPTES; Fluka, Milan, Italy), tetraethylorthosilicate (TEOS; Aldrich, Milan, Italy), hydrochloric acid at a 37% concentration (HCl, Carlo Erba, Milan, Italy), and ethanol (EtOH; Carlo Erba, Milan, Italy) were high-purity reagents and were used as received without further purification. Phenol red (PR; Aldrich, Milan, Italy) was used as received as pH-sensitive indicator. α,ω-Triethoxysilane-terminated poly(ethylene oxide) (called PEOSi in the following) was prepared by the bulk reaction of the corresponding α, ω-hydroxy-terminated poly(ethylene oxide) with ICPTES (molar ratio of 1:2). The reaction was carried out in a 50 mL glass flask equipped with a calcium chloride trap and under magnetic stirring at 120 °C for 3 h, following a reported procedure (Messori, 2004b).

3.2 Preparation of the sensitive matrix

PEOSi/SiO$_2$ hybrids were prepared by dissolving TEOS and PEOSi in EtOH at a concentration of about 60% (w/v), and then adding water (for the hydrolysis reaction) and HCl in the following molar ratios with respect to the ethoxide groups of both the PEOSi and TEOS: EtO-/H$_2$O/HCl 1/1/0.05. Then pH indicator PR was dissolved into the mixture (approx. 4 mg/50 mL) and left under stirring at room temperature for 15 min. The homogeneous indicator-doped mixture was finally heated at 60 °C for 15–30 min (in order to approach the gel point) before application of a drop onto the tip of the optical fiber.

The hybrid material was characterized by final organic:inorganic weight ratio of 8:2 (assuming the completion of the sol–gel reactions). Samples were coded as follows: PEO-PR where PEO refers to the polymer phase within the hybrid and PR indicates the pH indicator entrapped in the hybrid.

3.3 Preparation of the sensing probe

The polymer-silica hybrid matrix, doped with pH indicator, was prepared as above reported; after the thermal treatment at 60 °C for 15–30 min, condensation reactions involving ethoxysilane groups induced an increase in the viscosity of the solution, leading to a gel-like material. A drop of hybrid gel was then manually deposited onto the tip of the plastic optical fiber, and the optical fiber allowed to dry at atmospheric pressure and room temperature for 2 h and then cured at 105 °C for 24 h to increase the extent of the sol-gel reaction and to induce a strong interfacial connectivity between the sensing matrix and the POF. The dimension of the hybrid sensitive element was optimized looking for the best compromise between the rate of the response of the optical sensor and its physical and mechanical robustness. The final appearance of the sensing probe is reported in Fig. 2.

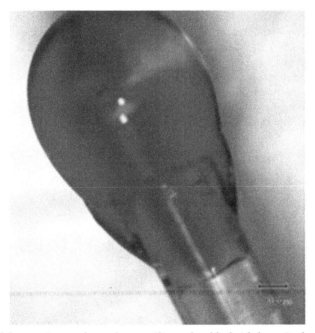

Fig. 2. Picture of the sensing probe: polymer-silica sol-gel hybrid deposited onto the tip of the POF.

4. Instruments and methods

In this section the instruments and methods used in the experimental activity are presented. To evaluate the performance of the developed probes a custom optical setup has been realized and characterized. Afterwards, ahead of the sensor characterization, the spectral properties of the sensing indicator have been determined.

4.1 Design and characterization of the optical setup

The instrumental setup realized to interrogate the sensing probe is shown in Fig. 3. It performs the illumination of the tip of the optical fiber and the collection of the reflected beam.

It consists of a white light source WLS (tungsten halogen lamp, HL-2000-HF, Avantes, Netherlands), two fiber-optic collimators (PAF-SMA-5-550, OFR, USA), a cube beam splitter (039-1130, Optosigma, USA), a glass electrode pH-meter (Eutech Instrument, pH 700 Netherlands) and a grating spectrometer OSA (PMA-11 C5966, Hamamatsu, Japan) equipped with a charge-coupled device detector. The beam, generated by the WLS, is delivered to the first collimator through a 0.5 m long multi-mode optical fiber (1 mm core diameter, NA 0,51, Fort Fibre Ottiche, Italy). The beam splitter, fed by the collimated light, transmits the beam to the second collimator that re-focuses it into the optical fiber sensor, whose sensitive tip is submerged into the testing solution. The glass electrode pH-meter monitors the testing solution.

The colored light reflected by the optical fiber sensor goes back over the optical fiber and reaches the entrance optical fiber bundle of the grating spectrometer OSA, after being deflected by the beam splitter. The measuring spectral range of our spectrometer covers the visible spectrum from 400 to 800 nm. The maximum resolution achievable is 0.487 nm; it has been computed dividing the spectral range by the number of acquisition spectral channels (400nm/820nm). In fact, according to the Rayleigh criterion, which points out how wide the spectral distance between two lines must be to allow their recognition as separate lines, the spectral width of the individual lines must be markedly smaller than their spacing. The theoretical limit computed for our spectrometer does not take into account channels crosstalk and other non-idealities.

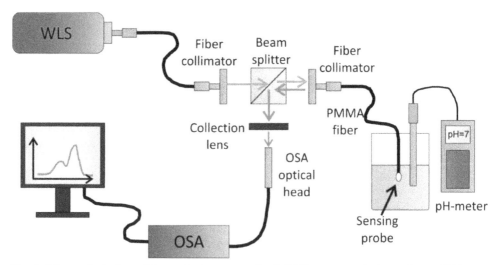

Fig. 3. Schematic design of the optical setup realized. WLS: tungsten halogen lamp, OSA: grating spectrometer.

In order to avoid the corruption of the spectral properties of the outgoing light because of the response of the measuring system and the transmission of the sensing POF, the spectral response of the interrogation system was measured using the experimental setup depicted in Fig. 4.

For this purpose, two measurements were performed; in particular, firstly, as shown in Fig. 4a, the light spectrum at the output of a 35 cm-long POF without sensing matrix was measured using spectrometer OSA. Afterwards, the spectrometer was reassembled as in Fig. 4b using the same 35 cm-long POF.

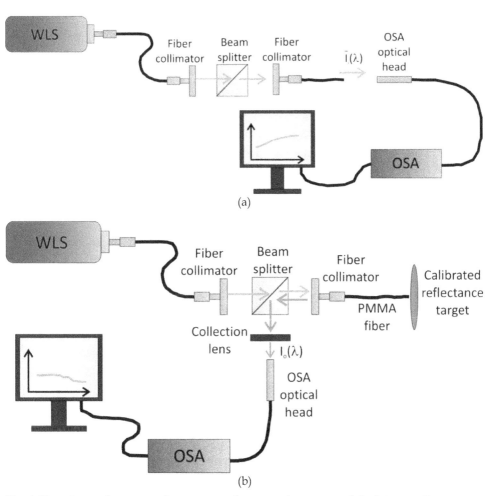

(a)

(b)

Fig. 4. Experimental setup used to measure the spectral response of the interrogation system. (a) The light spectrum at the output of a 35 cm-long POF without sensing matrix was measured. (b) Afterwards, the spectrometer was reassembled and in front of the POF a calibrated reflectance target acted as a diffuser.

A NIST (National Institute of Standards and Technology) calibrated reflectance target ($\Im$ =99%, Labsphere, USA) acted as a lambertian diffuser in front of the POF. This target has a spectrally flat (± 4%) reflectance over the UV-NIR spectrum. Spectrum $I_0(\lambda)$ acquired by OSA can be calculated as:

$$I_0(\lambda) = k \cdot \Im \cdot S_r(\lambda) \cdot \tilde{I}_i(\lambda) \tag{5}$$

where k is a coupling constant, $\Im$ the calibrated target reflectance assumed not dependent on the wavelength and $S_r(\lambda)$ the interrogation system response. Therefore, the ratio $\frac{I_0(\lambda)}{\tilde{I}_i(\lambda)}$ has been used to study the spectral trend of the interrogation system response.

In particular, the transmission features of the POF have been determined from the comparison between the spectrum of the halogen lamp WLS and $\tilde{I}_i(\lambda)$; as reported in literature (Feng, 2011), the POF exhibits an attenuation peak at about 750 nm.

4.2 Indicator optical characterization

Subsequent to the characterization of the optical properties of the interrogation system, in this section we are going to report the optical measurements performed to determine the spectral properties of the sensing indicator.

The absorbance spectra of phenol red in distilled water at different pH values were measured using a commercial spectrometer (Lambda 19 Spectrometer, Perkin Elmer USA); thus the absorption values at 560 nm were normalized and fitted by the Boltzmann (sigmoidal) function according to Eq. (6).

$$S = S_0 - \frac{a}{1 + e^{\frac{pH - pK}{b}}} \tag{6}$$

where a, b and S_0 are constants.

4.3 Sensors characterization

In our disposable sensor, the spectrum of the outgoing light $I_0(\lambda, pH)$ and thus the response of the sensor $\Re(\lambda, pH)$ is determined by two phenomena: the photon diffusion process in the hybrid-material drop applied to the optical fiber tip and the absorption of the indicator. Both these processes are pH-dependent; indeed, the size of the hybrid drop depends on the pH level of the solution and so, varying the pH, different photon optical paths are possible.

In order to characterize the sensing probe, measurements were performed at room temperature in standard pH buffers (Fisher Scientific) evaluating the response of the disposable sensor $\Re(\lambda, pH)$. According to the schematic representation in Fig. 5, the sensitive tip of the POF was immersed in 20 ml of sample solution; measurements were taken at 5 min after changing the pH level, monitored with the reference pH meter, which has a measuring uncertainty of 0.05 units. The system response $\Re(\lambda, pH)$ was calculated acquiring a reference spectrum $I_0(\lambda, pH_{ref})$ at pH_{ref}=10 and varying the pH of the solution from 9 to 4.

To evaluate the temporal response of the probe to a rapid change in pH, the pH value of the solution has been staircase decreased from 8 to 5 with unitary steps of duration 10 s each. The same protocol has been applied for the reverse direction of the staircase, i.e. from 5 to 8. Also during these measurements the testing solution was continuously monitored with the reference pH meter. The temporal response has been investigated firstly visually inspecting the color changes and then acquiring continuously the spectrum $I_0(\lambda, pH)$ and calculating $\Re(\lambda, pH)$ as a function of time.

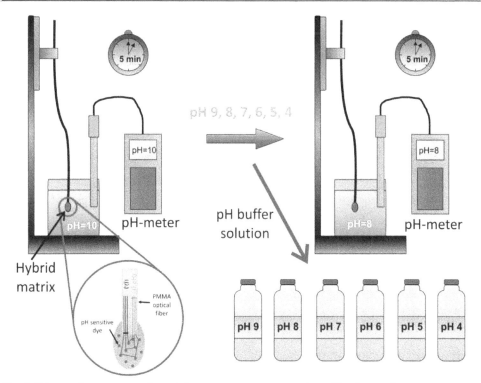

Fig. 5. Schematic representation of the measurement procedure. Firstly, a reference spectrum $I_0(\lambda, pH_{ref})$ at pH_{ref}=10 was acquired. Then, the pH has been changed from 9 to 4, using standard pH buffers and the response of the disposable sensor has been taken at 5 min after changing the pH level, monitored with the reference pH meter.

5. Results and discussion

The optical properties of the interrogation system have been determined according to the measurement procedures reported in section 4.1. Spectra $I_0(\lambda)$, $\tilde{I}_i(\lambda)$ and their ratio are shown in Fig. 6 together with the emission spectrum of the halogen lamp WLS. Other distortions of the halogen lamp spectrum are due to the transmission properties of the optical components used in the interrogation setup.

Figure 6b shows the ratio $\frac{I_0(\lambda)}{\tilde{I}_i(\lambda)}$ that, according to Eq. (5), is proportional to the spectral response of the interrogation system. The spectral region of interest in our application is 500-600 nm, i.e. green-yellow-red; since the response of the interrogation system is satisfactorily flat in this interval, as shown in Fig. 6b, the information of interest is not corrupted.

The spectral properties of the sensing indicator have been measured according to the procedure reported in section 4.2. The absorbance spectra of phenol red at different pH values are shown in Fig. 7; the maximum of all the spectra is well visible at the wavelength of about 560 nm. According to literature (Wang, 2003), at this wavelength the phenol red

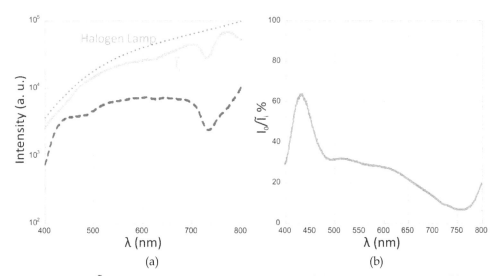

Fig. 6. Spectra of $\tilde{I}_i$ (λ) and $I_o(\lambda)$ (a), and spectral response of the interrogation system (b). In (a), the upper dashed blue line represents the emission spectrum of the halogen lamp WLS. The spectral region of interest for our application is 500-600 nm.

molecules exhibit a very low absorption that can be significantly enhanced by increasing the pH of the solution. Note that the absorbance at 560 nm showed a sharp transaction when the pH of the solution changed from 5 to 8.

As shown in Fig. 8, the point of inflection (sigmoidal fitting function Eq. (6) equal to 0.5) corresponds to the apparent pKa value of 7.9. This value is in good agreement with the pKa of phenol red in aqueous solution reported in literature (Budavari, 1989).

The sensor characteristics have been measured according to the procedure reported in section 4.3. As shown in Fig. 9, a visual inspection of the probe at two different pH values of the solution underlines the swelling described in the previous sections. The spectra $\Delta I_0(\lambda, pH) = I_0(\lambda, pH) - I_0(\lambda, pH_{ref})$ acquired at different values of pH are shown in Fig. 10.

Observing Fig. 10, it is clear that the swelling/shrinking processes induce a change in the amplitude of the whole spectrum. As far as these phenomena are concerned, parameters A and S were considered to quantify the response of the sensor. In particular, as shown in Fig. 11, given a typical spectrum $\Delta I_0(\lambda)$, parameter A and S were calculated as follow: (i) the two maxima $M_1 = (\lambda_1, I_1)$, $M_2 = (\lambda_2, I_2)$ and the minimum $m = (\lambda_3, I_3)$ of $\Delta I_0(\lambda)$ have been calculated using a standard Matlab© routine; (ii) the equation of the line joining M_1 and M_2 has been calculated as $L(\lambda) = \frac{I_2 - I_1}{\lambda_2 - \lambda_1}(\lambda - \lambda_1)$; (iii) A has been defined as $L(\lambda_3) - I_3$ and S as $L(\lambda_3)$.

As shown in Fig. 11, parameter S is correlated to the spectrum amplitude and thus to swelling/shrinking processes while A is strictly related to the absorption (at about $\lambda = 560$ nm) performed by the indicator. In Fig. 12, parameters A and S were calculated for each spectrum acquired and the normalized results are plotted against pH.

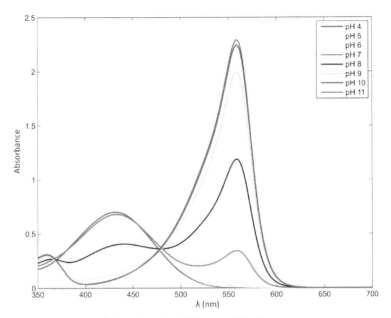

Fig. 7. Absorbance spectra of phenol red at different pH values.

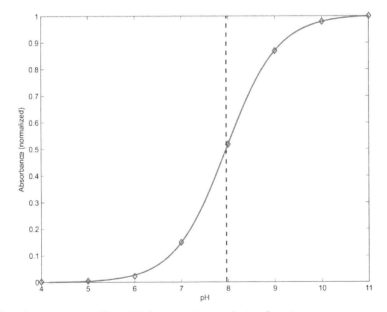

Fig. 8. Absorbance versus pH and Boltzmann interpolation function.

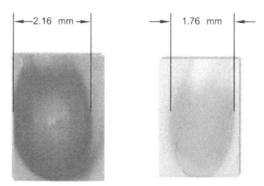

Fig. 9. Two shapes appearance of the probe PEO8000–8:2-PR after 5 min of immersion in basic (left) and acid (right) solution.

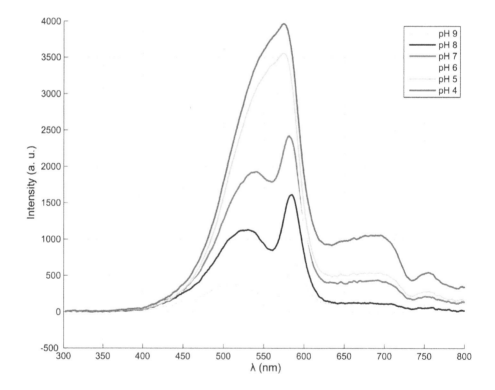

Fig. 10. Spectra of $\Delta I_0(\lambda)$ acquired at different values of pH. The light absorption of phenol red is well visible at the wavelength of 560 nm. Swelling/shrinking processes induce a change in the amplitude of the whole spectrum.

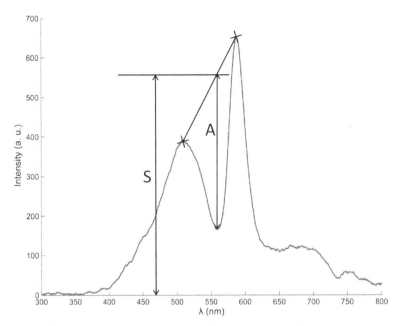

Fig. 11. Graphical definition of the two characteristic parameters. Parameter A is strictly related to the absorption performed by the indicator, whereas S is correlated to the swelling/shrinking processes.

Parameter A exhibits a sharp transaction changing the pH of the solution from 5 to 8. The experimental data was fitted by the Boltzmann (sigmoidal) function according to Eq. (6). The point of inflection (sigmoidal fitting function equal to 0.5) corresponds to the apparent pKa value $pK'_a = 6.8$. This value is lower than the measured value of pKa=7.9 of phenol red in aqueous solution (Fig. 8). However, the pKa value can be shifted to lower pH with increasing ionic strength as observed by Holobar et al. (Holobar, 1993), and this phenomenon could occur also in our hybrid matrix.

An increase in the value of pH determines a higher swelling of the hybrid drop, as shown in Fig. 9, which leads to an increase in the distance between the optical fiber tip and the probe-liquid interface and a consequent decrease in the mean optical path of the photons through the probe. As a consequence, according to Eq. (3), also the sensor response diminished. Coherently, as shown in Fig. 12b, parameter S decreased linearly with the pH of the solution. The linear fit of the experimental data exhibits a slope of 0.17.

The response time of the probe PEO8000-8:2-PR was measured. A trade-off between the response time and the indicator leaching together with the adhesion between the sensitive matrix and the POF is required. In fact, high molecular weight PEO improves the adhesion and diminishes the indicator leaching, but causes long response time of the sensor. In general there are two different time constant; in fact, when pH is changed from 10 to 3, the sensor was found to reach 90% of the regime signal intensity (τ_{90}) between 1 and 2 seconds, while changing the pH from a low value to a high one resulted in longer response times that can be quantify in the order of some minutes, i.e. the kinetics depends on the direction of the

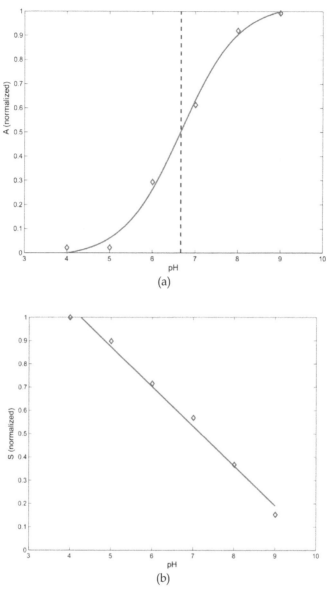

(a)

(b)

Fig. 12. Normalized A and S as a function of the pH value. Absorption parameter A is fitted by the Boltzmann function (a). Linear fit of the normalized S exhibits a slope of 0.17 (b).

pH change (the 90% of the regime signal intensity was not reached in the time scale of the experiment, i.e. 10 seconds per step). This agrees with other experimental observations reported in literature (Ismail, 2002; Badini, 1995). Empirical quantification of this sensor behavior is shown in Fig. 13.

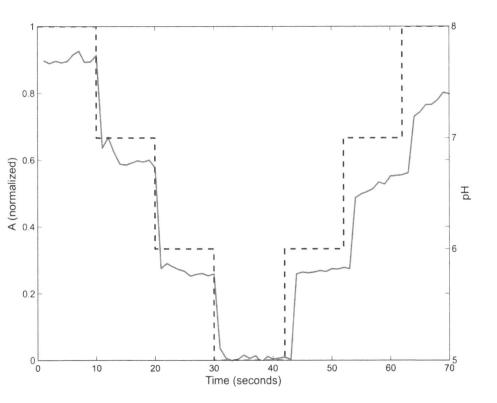

Fig. 13. Normalized A as a function of time for different values of pH. The pH value of the solution has been staircase changed from 8 to 5 and then backs from 5 to 8 with unitary steps of duration 10 s each.

In Fig. 13, the parameter A (blue line) is represented as a function of time together with the value of the reference pH (dashed line).

We think that the slow response time observed changing the pH from 5 to 8 is strictly associated with the swelling/shrinking kinetics of the sensitive drop. Finally, the kinetics of swelling/shrinking has shown to be dependent on the direction of the pH change. This complex mechanism meddles in the photons propagation in the sensitive drop and thus in the response of the probe.

6. Characteristics of the developed pH sensor

The mean lifetime of disposable sensors, checked in the continuous mode operation at a pH level in the middle of the range, is two days. The reusability, tested with a series of probes checked once a day at a pH level in the middle of the range and preserved dry, has a mean life of one month. Table 1 shows these and other analytical parameters of the sensor.

Parameter	Value
Optical interrogation wavelength	560 nm
Probe diameter	2 mm (typ.)
Measuring pH Range	5-8
Sensitivity	0.5 a.u./pH unit (max.)
Response time pH 8 to pH 5	2 s (max.)
Response time pH 5 to pH 8	4 min (typ.)
Reusability	1 month
Mean lifetime	2 days
Cost (few probes)	less than 0.7 Euro

Table 1. Characteristics of the developed disposable pH probe.

7. Conclusions

We present a disposable sensor based on a low-cost plastic optical fiber that covers an important pH range for physiological applications, e.g. in medicine and biotechnology, with excellent sensitivity. Some other advantages of the proposed sensor are: simple realization process, small size, high resistance to aggressive environments and electrical isolation.

The sensor characterization has been performed using a simple optical setup based on a single wavelength measurement. Nevertheless according to Lei (Lei, 2010), a ratiometric approach could improve the sensing overall performance. To this purpose, the isosbestic wavelength, i.e. 480 nm, highlighted in the spectra shown in Fig. 7, could be conveniently exploited.

For practical applications, swelling/shrinking processes could make not trivial the extraction of the pH information and deteriorate the sensor response time. Nevertheless, this phenomenon can be practically cancelled minimizing the size of the sensing element. Large drop volume allows to make the probe fabrication simple and reproducible and to obtain high signal-to-noise ratio but emphasizes the swelling/shrinking effects. Depending on the application, the trade-off among these aspects can establish the optimal drop size.

8. References

Alvarado-Mendez E., Rojas-Laguna R., Andrade-Lucio J.A., Hernandez-Cruz D., Lessard R.A. & Avina-Cervantes J.G., (2005). Design and characterization of pH sensor based on sol–gel silica layer on plastic optical fiber. Sens. Actuator B-Chem.: Vol. 106, pp. 518–522, ISSN 0925-4005

Aubonnet S., Barry H.F., von Bueltzingsloewen C., Sabbatie J.M. & MacCraith B.D., (2003). Photo-patternable optical chemical sensors based on hybrid sol-gel materials. Electron. Lett., Vol.39, No.12, pp.913-914, ISSN 0013-5194

Bacci M., Baldini F.& Bacci S. (1991). Spectroscopic behavior of acid-base indicators after immobilization on glass supports. Appl. Spectrosc., Vol.45, No.9, pp.1508-1515, ISSN 0003-7028

Badini G.E., Grattan K.T.V. & Tseung A.C.C. (1995). Impregnation of a pH-sensitive dye into sol-gels for fibre optic chemical sensors. Analyst, Vol.120, No.4, pp. 1025-1028, ISSN 0003-2654

Baldini F., Giannetti A., Mencaglia A.A. & C. Trono. (2008). Fiber-optic sensors for biomedical applications. Curr. Anal. Chem., Vol.4, pp. 378-390; ISSN 1573-4110

Barnard S.M.& Walt D.R. (1991). A fibre-optic chemical sensor with discrete sensing sites. Nature, Vol.353, pp. 338-340, ISSN 0028-0836

Bashir R., (2002), Micromechanical cantilever as an ultrasensitive pH microsensor. Appl. Phys. Lett. , 81, 3091-3093; ISSN 0003-6951

Ben-David O., Shafir E., Gilath I., Prior Y. & Avnir D. (1997). Simple absorption optical fiber pH sensor based on doped sol-gel cladding material. Chem. Mat.,Vol.9, pp. 2255-2257, ISSN 0897-4756

Brinker C.J. & Scherer G.W. (1990). Sol–gel science. The physics and chemistry of sol–gel processing. Academic Press, ISBN: 978-0121349707, New York

Bronk K. & Walt D.R. (1994). Fabrication of patterned sensors arrays with aryl azides on a polymer-coated imaging optical fiber bundle. Anal. Chem, Vol.66, No.20, pp. 3519-3520, ISSN 1520-6882

Budavari S., O'Neil M.J., Smith A. & Heckelman P.E. (1989). The Merck Index, 11th ed. Merck & Co., ISBN: 0-911910-28-X, Inc. Rahway, NJ

Cagnol F., Grosso D. & Sanchez C. (2004). A general one-pot process leading to highly functionalised ordered mesoporous silica films. Chem. Commun., pp. 1742-1743, ISSN 1359-7345

Cao W. & Duan Y. (2005). Optical fiber-based evanescent ammonia sensor. Sens. Actuator B-Chem., Vol.110, No.2, pp. 252-259, ISSN 0925-4005

Cjlakovic M., Lobnik A. & Werner T. (2002). Stability of new optical pH sensing material based on cross-linked poly(vinyl alcohol) copolymer. Anal. Chim. Acta, Vol.455, pp. 207-213, ISSN 0003-2670

de Marcos S. & Wolfbeis O.S. (1996). Optical sensing of pH based on polypyrrole films. Anal. Chim. Acta, Vol.334, pp. 149-153, ISSN 0003-2670.

Dong S., Luo M., Peng G. & Cheng W. (2008). Broad range pH sensor based on sol–gel entrapped indicators on fiber optic. Sens. Actuator B-Chem., Vol. 129, pp. 94–98, ISSN 0925-4005

Eckert H. & Ward M. (Eds.). (2001) Nanostructured and functional hybrid organic–inorganic materials. Chem. Mat., Vol. 13, No. 10. ISSN 0897-4756

Estella J., Echeverria J.C., Laguna M. & Julian J. (2007). Silica xerogels of tailored porosity as support matrix for optical chemical sensors. Simultaneous effect of pH, ethanol:TEOS and water:TEOS molar ratios, and synthesis temperature on gelation time, and textural and structural properties. J. Non-Cryst. Solids, Vol.353, No.3, pp. 286 294, ISSN 0022-3093

Fabbri P., Leonelli C., Messori M., Pilati F., Toselli M., Veronesi, P., Morlat-Therias S., Rivaton A. & Gardette J.L. (2008). Improvement of the surface properties of polycarbonate by organic-inorganic hybrid coatings. J. Appl. Polym. Sci., Vol. 108, No. 3, pp. 1426-1436. ISSN 0021-8995

Fabbri P., Pilati F., Rovati L., McKenzie R. & Mijovic J. (2011). Poly(ethylene oxide)-silica hybrids entrapping sensitive dyes for biomedical optical sensors: molecular dynamics and optical response. Opt. Mater., 33(8), 1362-1369 ISSN 0925-3467

Feng C. & Zhongyang R. (2011). Research on Measurement of POF Attenuation Spectrum. Intelligent Computation Technology and Automation (ICICTA), 2011 International Conference on , Vol.2, pp.672-674.

Ferguson J.A., Healey B.G., Bronk K.S., Barnard S.M. & Walt D.R. (1997). Simultaneous monitoring of pH, CO_2 and O_2 using an optical imaging fiber. Anal. Chem, Vol.340, pp 123-131; ISSN 1520-6882

Fritz J., Baller M.K., Lang H.P., Strunz T., Meyer E., Guntherodt H.J., Delamarche E., Gerber Ch. & Gimzewski J.K. (2000). Stress at the solid–liquid interface of self-assembled monolayers on gold investigated with a nanomechanical sensor. Langmuir, Vol.16, (November 2000), pp 9694-9696, ISSN 0743-7463

Goicoechea J., Zamarreno C.R., Matias I.R. & Arregui F.J. (2009). Utilization of white light interferometry in pH sensing applications by mean of the fabrication of nanostructured cavities. Sens. Actuator B-Chem., Vol.B138, pp 613-618, ISSN 0925-4005

Grosso D., Cagnol F., Soler-Illia G.J., Crepaldi E.L., Amenitsch H., Brunet-Bruneau A., Bourgeois A. & Sanchez C. (2004). Fundamentals of mesostructuring through evaporation-induced self-assembly. Adv. Funct. Mater. Vol.14, No.4, pp. 309-322, ISSN 1616-301X

Grummt U.W., Pron A., Zagorska M. & Lefrant S. (1997). Polyaniline based optical pH sensor. Anal. Chim. Acta, Vol.357, pp. 253-259; ISSN 0003-2670

Gupta B.D. & Sharma S. (1998). A long-range fiber optic pH sensor prepared by dye doped sol–gel immobilization technique. Opt. Commun., Vol. 154, pp. 282-284, ISSN 0030-4018

Healey B.G. & Walt D.R. (1997). Fast temporal response fiber-optic chemical sensors based on the photodeposition of micrometer-scale polymer arrays. Anal. Chem, Vol.69, No.11, pp. 2213-2216; ISSN 1520-6882

Holobar A., Weigl B.H., Trettnak W., Benes R., Lehmann H., Rodriguez N.V., Wollschlager A., O'Leary P., Raspor P. & Wolfbeis O.S. (1993). Experimental results on an optical pH measurement system for bioreactors. Sens. Actuator B-Chem., Vol.11, No.1-3, pp. 425-430, ISSN 0925-4005

Igarashi S., Kuvae K. & Yotsuyanagi T. (1994). Optical pH sensor of electrostatically immobilized porphyrin on the surface of sulfonated-polystyrene. Anal. Sci., Vol.10, (October 1994), pp. 821-822, ISSN 0910-6340

Ismail F., Malins C. & Goddard N.J. (2002). Alkali treatment of dye-doped sol-gel glass films for rapid optical pH sensing. Analyst, Vol.127, No.2, pp. 253-257, ISSN 0003-2654.

Kickelkick G. (2006). Hybrid materials. Synthesis, characterization, and application. Wiley-VCH, , ISBN 3527312994, New York

Korostynska O., Arshak K., Gill E. & Arshak A. (2007), Review on state of the art in polymer based pH sensors. Sensors, Vol.7, pp 3027-3042, ISSN 1424-8220

Kowada Y., Ozeki T & Minami T. (2005). Preparation of silica-gel film with pH indicators by the sol-gel method. J. Sol-Gel Sci. and Techn., Vol.33, pp. 175-185, ISSN 0928-0707

Lee S.T., Gin J., Nampoori V.P.N., Vallabhan C.P.G., Unnikrishnan N.V. & Radhakrishnan P. (2001). A sensitive fiber optic pH sensor using multiple sol–gel coatings. J. Opt. A-Pure Appl. Opt., Vol. 3, pp. 355–359, ISSN 1464-4258

Lei J., Wang L. & Zhang J. (2010). Ratiometric pH sensor based on mesoporous silica nanoparticles and Förster resonance energy transfer. Chem. Commun., Vol. 46, pp 8445-8447, ISSN 1359-7345

Lin J. (2000). Recent development and applications of optical and fiber-optic pH sensors. Trends Anal. Chem., Vol.19, No. 9, pp 541-552, ISSN 0167-2940

Lin J. & Brown C.W. (1997). Sol-gel glass as a matrix for chemical and biochemical sensing. Trac-Trends Anal. Chem.Vol.16, No.4, pp. 200-211, ISSN 0167-2940

Lobnik A., Oehme I., Murkovic I. & Wolfbeis O.S. (1998). pH optical sensors based on sol-gels: chemical doping versus covalent immobilization. Anal. Chim. Acta, Vol.367, pp. 159-165, ISSN 0003-2670

MacGraith B.D., Ruddy V., Potter C., O'Kelly B. & McGilp J.F. (1991). Optical waveguide sensor using evanescent wave excitation of fluorescent dye in sol–gel glass. Electron. Lett., Vol. 27, pp. 1247–1248, ISSN 0013-5194

McDonagh C., Burke C.S. & MacCraith B. D. (2008). Optical chemical sensors. Chem. Rev., Vol.108, pp. 400-422, ISSN 0009-2665

Messori M., Toselli M., Pilati F., Fabbri E., Fabbri P., Busoli S., Pasquali L. & Nannarone S. (2003). Flame retarding poly(methyl methacrylate) with nanostructured organic-inorganic hybrids coatings. Polymer, Vol. 44, No. 16, pp. 4463-4470, ISSN: 0032-3861

Messori M., Toselli M., Pilati F., Fabbri E., Fabbri P. & Busoli S. (2004a). Poly(caprolactone)/silica organic-inorganic hybrids as protective coatings for poly(methyl methacrylate) substrates. Surf. Coat. Int., Vol. 86, No. 3, pp. 183-186, ISSN 1754-0925

Messori M., Toselli M., Pilati F., Fabbri E., Fabbri P., Pasquali L. & Nannarone S. (2004b). Prevention of plasticizer leaching from PVC medical devices by using organic-inorganic hybrid coatings. Polymer Vol. 45, pp. 805–813, ISSN 0032-3861

Miled O.B., H.B. Ouada & J. Livage (2002). pH sensor based on a detection sol–gel layer onto optical fiber. Mater. Sci. Eng. C-Mater. Biol. Appl., Vol. 21, pp. 183–188, ISSN 0928-4931

Pringsheim E., Terpetschnig E. & Wolfbeis O.S. (1997). Optical sensing of pH using thin films of substituted polyaniline. Anal. Chim. Acta, Vol.357, pp. 247-252, ISSN 0003-2670

Rayss J. & Sudolski S. (2002). Ion adsorption in the porous sol–gel silica layer in the fiber-optic pH sensor. Sens. Actuator B-Chem., Vol. 87, pp. 397–405, ISSN 0925-4005

Rovati L., Fabbri P., Ferrari L. & Pilati F. (2011). Construction and evaluation of a disposable pH sensor based on a large core plastic optical fiber. Rev. Sci. Instrum., Vol. 82, p. 023106, ISSN 0034-6748

Ruan C., Zeng K. & Grimes C.A. (2003). A mass-sensitive pH sensor based on a stimuli-responsive polymer. Anal. Chim. Acta, Vol.497, pp 123-131, ISSN 0003-2670

Sanchez C., Shea K.J. & Kitagawa S. (2011). Recent progress in hybrid materials science. Chem. Soc. Rev., Vol. 40, pp. 471-472, ISSN. 0306-0012

Schmidt H. (2000). Sol-gel derived nanoparticles as inorganic phases in polymer-type matrices. Macromol. Symp., Vol. 159, pp. 43–55, ISSN 1521-3900

Seki A., Katakura H., Kai T., Iga M. & Watanabe K. (2007). A hetero-core structured fiber optic pH sensor. Anal. Chim. Acta, Vol 582, No.1, pp. 154-157, ISSN 0003-2670

Sharma N.K. & Gupta B.D. (2004). Fabrication and characterization of a fiber-optic pH sensor for the pH range 2 to 13. Fiber Integrated Opt., Vol. 23, pp. 327–335 ISSN 0146-8030

Shi Y. & Seliskar C.J. (1997). Optically transparent polyelectrolyte-silica composite materials: preparation, characterization, and application in optical chemical sensing. Chem. Mater., Vol.9, pp. 821-829, ISSN 0897-4756

Song A., Parus S. & Kopelman R. (1997). High-performance fiber-optic pH microsensors for practical physiological measurements using a dual-emission sensitive dye. Anal. Chem, Vol.69, No.5, (March 1997), p. 863867, ISSN 1520-6882

Sotomayor M.D.P.T., De Paoli M.A. & Oliveira W.A.D. (1997). Fiber-optic pH sensor based on poly(o-methoxyaniline). Anal. Chim. Acta, Vol.353, pp 275-280, ISSN 0003-2670

Splinter R., Brett A. & Hooper (2006). An introduction to biomedical optics. Taylor & Francis Ed, ISBN: 0750309385 9780750309387, New York

Talaie A., Lee J.Y., Jang J., Romagnoli J.A. & Taguchi T. (2000). Dynamic sensing using intelligent composite: an investigation to development of new pH sensors and electrocromic devices. Thin Solid Films, Vol. 363, pp. 163-166, ISSN 0040-6090

Toselli M., Marini M., Fabbri P., Messori M., & Pilati F. (2007). Sol-gel derived hybrid coatings for the improvement of scratch resistance of polyethylene. : J. Sol-Gel Sci. Technol., Vol. 43, No. 1, pp. 73-83, ISSN 0928-0707

Wang E., Chow K.F., Kwan V., Chin T., Wong C. & Bocarsly A. (2003). Fast and long term optical sensors for pH based on sol–gels. Anal. Chim. Acta, Vol. 495, No.1-2, pp. 45-50, ISSN 0003-2670

Wen J. & Wilkes G.L. (1996). Organic/inorganic hybrid network materials by the sol-gel approach. Chem. Mater., Vol.8, No.8, pp. 1667-1681, ISSN 0897-4756

Wolfbeis O.S. (2004). Fiber-optic chemical sensors and biosensors. Anal. Chem., Vol.76, No.12, pp. 3269-3284, ISSN 1520-6882

Wolfbeis O.S. (2008). Fiber-optic chemical sensors and biosensors. Anal. Chem., Vol.80, No.12, pp. 4269-4283, ISSN 1520-6882

Wu Z., Jiang Y., Xiang H., You L. (2006). Understanding the mechanisms of reaction and release of acid–base indicators entrapped in hybrid gels. J. Non-Cryst. Solids, Vol. 352, pp. 5498–5507, ISSN 0022-3093

Yuqing M., Jianrong C., & Keming F. (2005). New technology for the detection of pH. J. Biochem. and Biophys. Methods, Vol.63, pp. 1-9, ISSN 0165-022X

Zamarreno C.R., Hernaez M., Del Villar I., Fernandez-Valdivielso C., Arregui F.J. & Matias I.R. (2010). Optical fiber pH sensor fabrication by means of Indium tin oxide coated optical fiber refractometers. Phys. Status Solidi C: Current Topics in Solid State Phys., Vol.7, No. 11-12, pp. 2705-2707, ISSN 1862-6351

Zauner A., Bullock R., Di X. & Young H.F. (1995). Brain oxygen, CO_2, pH and temperature monitoring: evaluation in the feline brain. Neurosurgery, Vol.37, No.6, pp. 1176-1177, ISSN 0148-396X

8

High-Sensitivity Detection of Bioluminescence at an Optical Fiber End for an ATP Sensor

Masataka Iinuma, Yasuyuki Ushio, Akio Kuroda and Yutaka Kadoya
Graduate School of Advanced Sciences of Matter, Hiroshima University
Japan

1. Introduction

In biological studies, the luminescence from fluorescent proteins or luminescent enzymes is widely used for monitoring a change of environment at a cell. Biomolecules used for this probing, such as Green Fluorescence Protein(GFP) or luciferase molecules can respond to the existence of specific molecules or ions and subsequently emit a photon. The detection of a specific molecule can then be confirmed by detecting the emitted photons efficiently with a photon detector. A highly efficient detection of the luminescence is normally essential to a high sensitivity to the specific molecules or ions. An improvement of the sensitivity can upgrade the capability of detection in a low concentration of sample solution. Therefore, there are many efforts to improve the efficiency of the collection of emitted photons and of the optical coupling to the photon detector (Yotter, 2004).

A straightforward approach is to directly detect the luminescence from the whole sample solution in a test tube as shown in Fig. 1 (a). However, to realize high efficiency detection, this method needs a single photon detector with a wide photon-sensitive area, which is ideally larger than the photon-emission area in the test tube. The reason is that it is difficult to image incoherent light such as natural light to a smaller area than the emission area. Here, we are introducing an alternative method, where the luminescent biomolecules are immobilized at an optical fiber end and the luminescence is detected by a photon detector which is optically coupled to the other optical fiber end. The sketch of the optical fiber-based systems is shown in Fig. 1 (b). This method has been investigated for application to a fiberoptic biosensor, which is constructed by immobilizing either an enzyme or an antibody. A review is given in (Arnold, 1991), (Blum & Gautier, 1991).

This method has three merits. The first one is to permit a local detection within the sample solution, because the optical fiber end functions as a needle-like probe in the solution. The second one is that the detection scheme does not require that the photon detector is very close to the sample solution. This feature makes it easier to mount the sensing parts in integrated bioengineering, such as μ-TAS. The third merit is that single photon detectors with a small sensitive area can be used, because the photon-emission area, which is almost identical to the cross section of the core part in the optical fiber, is small. In general, single photon detectors have lower dark counts for smaller sensitive area. Low dark counts are very significant, because it essentially gives the upper limit of the sensitivity of photon detection. Recently, single photon detectors using avalanche photodiodes (APDs) have become widely available with good performance, but their sensitive area is small and has a typical size of 0.1 mm. The

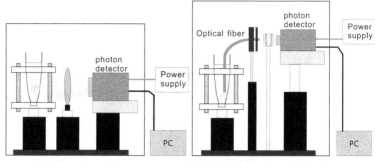

(a) Direct detection system   (b) Optical fiber-based detection system

Fig. 1. Schematic figure of two detection systems

luminescence detection by the optical fiber-based system allows us to fully use the merits of compactness, high quantum efficiency, and low noise of these APD detectors.

We have built a detection system of bioluminescence at an optical fiber end and investigated the sensitivity of Adenosine triphosphate (ATP) detection by using an APD-type photon detector (Iinuma et al., 2009). ATP is a good indicator of biochemical reaction or life activity, since ATP is considered as the universal currency of biological energy for all living things. Therefore, there are a lot of efforts to develop ATP-sensing techniques for compact and efficient ATP detection (Stanley, 1992), (Andreotti & Berthold, 1999), (Gourine et al., 2005). In particular, high-sensitivity detection of ATP can indicate the existence of microorganisms even in low numbers. Thus, a compact and simple detection system with extremely high sensitivity is very desirable.

One powerful method for highly sensitive ATP detection is to use the chemical reaction involved in the bioluminescence, the luciferin-luciferase reaction (Fraga, 2008). In this reaction, after one ATP molecule and one luciferin molecule are bound to one luciferase molecule, the luciferin molecule is oxidized using the energy of ATP. As a consequence, one photon is emitted during the transition from the excited state to the ground state of the oxidized luciferin molecule bound to the luciferase molecule. The emission of one photon indicates the use of the energy of one ATP molecule. In the method using the luciferin-luciferase reaction, the efficient detection of the bioluminescence is essential for high-sensitivity detection of ATP.

Since the oxidation of luciferin is catalyzed by the enzyme luciferase, the immobilization of luciferase molcules on solid probes of various sizes allows highly sensitive and local measurements of ATP. Three types of immobilization have been used: firstly attachment to the cell surface (Nakamura et al., 2006), secondly attachment to small particles, such as nanoparticles (Konno et al., 2006) and glass beads (Lee et al., 1977), thirdly attachment to extended objects with a size in the centimeter range, such as strips (Blum et al., 1984), (Ribeiro et al., 1998) and films (Worsfold & Nabi, 1986). For the ATP-detection on an intermediate scale below 1 millimeter, a fiberoptic probe employing immobilized luciferase (Blum & Gautier, 1991) as well as microchips (Tanii et al., 2001), (Tsuboi et al., 2007), can be utilized. The detection system of bioluminescence at an optical fiber end can achieve local detection of ATP within the sample solution. Realization of high sensitivity potentially provides the local detection of extremely low number of microorganisms. Thus, it is desirable to construct a highly efficient detection system of the bioluminescence at an optical fiber end and to evaluate the detection limit with the system.

The rest of this chapter is organized as follows. In sec. 2, we describe a concept for the construction of the optical fiber-based system for efficient detection of a fluorescence at the optical fiber end. In sec. 3, we show how to optimize the parameters of the optical fiber and the coupling optics so as to realize high photon-collection efficiency. In sec. 4, we describe the application of the constructed detection system to ATP sensing. By immobilizing luciferase molecules at the optical fiber end, the bioluminescence by luciferin-luciferase reaction can be detected using the optical fiber-based system. We evaluated the sensitivity of ATP with this system. Sec. 5 summarizes present results and problems.

2. General concept for construction of the optical fiber-based system

For a luminescence detection system using an optical fiber with a core diameter ϕ_0 and a numerical aperture NA_0, a collection efficiency of the luminescence η at the optical fiber end depends only on NA_0 as shown in Fig. 2. From a simple calculation based on the solid angle

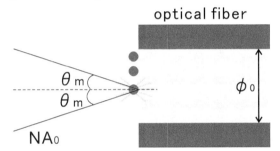

Fig. 2. Fluorescence at the optical fiber end. θ_m is a maximum opening angle for light propagation in the optical fiber.

with a maximum opening angle θ_m, $\eta(NA_0)$ can be expressed as,

$$\eta(NA_0) = \frac{1}{2}(1 - \cos\theta_m)$$

$$- \frac{1}{2}\left[1 - \sqrt{1 - \left(\frac{NA_0}{n_w}\right)}\right],\tag{1}$$

where n_w is the refraction index of the substance surrounding the optical fiber end. In immersing the optical fiber end into water, its value should be identical to the value of water, which is about 1.33. Fig. 3 shows the calculated values of the collection efficiency $\eta(NA_0)$ as a function of NA_0 using Eq. (1) at $n_w = 1.33$. One can easily see that $\eta(NA_0)$ increases with NA_0.

In the following, let us consider the situation where the other optical fiber end is optically coupled to a photon detector with a circular sensitive area having a diameter ϕ_3 and a numerical aperture NA_3. The coupling efficiency ϵ between the optical fiber end and the photon detector depends on ϕ_0, NA_0 of the optical fiber and ϕ_3, NA_3 of the photon detector

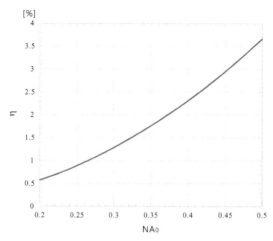

Fig. 3. Calculated values of collection efficiency as a function of NA_0 at $n_w=1.33$

used. Since the number of emitted photons is proportional to the square of ϕ_0 and $\eta(NA_0)$ increases with NA_0 as expressed by Eq. (1), the number of transmitted photons to the other optical fiber end is proportional to the square of ϕ_0 and $\eta(NA_0)$. On the other hand, the coupling efficiency ϵ generally decreases as ϕ_0 or NA_0 increases. Thus, we can define the following formula for a figure of merit (FOM) and optimize ϕ_0, NA_0 and parameters of the coupling optics x_i to maximize this FOM:

$$FOM = \phi_0^2 \cdot \eta(NA_0) \cdot \epsilon(\phi_0, NA_0, x_i; \phi_3, NA_3) \tag{2}$$

It should be noted that ϵ can ideally be 100 % under the conditions of $\phi_0 \leq \phi_3$ and $NA_0 \leq NA_3$. In many cases, the condition $NA_0 \leq NA_3$ is satisfied when using typical photon detectors. Thus, under the condition of $NA_0 \leq NA_3$, we can classify two cases: case (1) is $\phi_0 \leq \phi_3$ and case (2) is $\phi_0 > \phi_3$. In case (1), ϵ is constant and can ideally be 100 % and the total detection efficiency is limited only by $\eta(NA_0)$. Therefore, optimization of the coupling optics is not necessary. The conditions $\phi_0 = \phi_3$ and $NA_0 = NA_3$ both maximize the FOM and the sensitivity becomes highest. In case (2), however, the optimization of ϕ_0, NA_0, and the parameters x_i for a design of the coupling optics are necessary for given values of NA_3 and ϕ_3, because ϵ decreases as ϕ_0 or NA_0 increases.

3. Design of coupling optics

3.1 Photon detectors

Photon detectors generally have two significant factors contributing to the sensitivity of detection for weak light: the efficiency and the dark counts of the photodetector. A cooled APD which can detect for single photons is mostly used because of its very low dark counts. The sensitive area must be very small for realizing a large reduction of the dark counts, but the quantum efficiency is several times larger than that of a photomultiplier tube(PMT). Furthermore, the APD has the useful characteristics of compactness, easy operation, and durability in comparison with a PMT. To construct an optical fiber-based system with high sensitivity, we chose an APD-type photon counting module (SPCM-AQR-14) manufactured by Perkin Elmer Co. Ltd., which has a quantum efficiency η_{qe} of 55 % at 550 nm and dark counts of about 100 s^{-1}. The APD has a circular sensitive area, where ϕ_3 is 0.175 mm and NA_3 is 0.78, as calculated from the geometrical structure between the sensitive area and the photon detection window. If we use an optical fiber with $\phi_0 > \phi_3$ to increase the number of emitted photons, it is necessary to optimize ϕ_0, NA_0, and to design the coupling optics for maximal sensitivity.

3.2 Design concept and procedure

We can consider the coupling optics between the optical fiber end and the APD as an optical system imaging a light source with NA_0 and ϕ_0 onto the APD with NA_3 and ϕ_3. The basic design of the coupling optics is shown in Fig. 4.

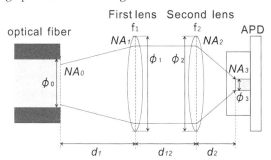

Fig. 4. Design of the coupling optics between the fiber output and the APD

Among the parameters shown in Fig. 4, we firstly determine the parameters of optical components, the focal length f_1 and the numerical aperture NA_1 of the first lens, the focal length f_2 and the numerical aperture NA_2 of the second lens, and the distance d_{12} between the first lens and the second lens. The second lens is selected among available lenses to make NA_2 as large as possible while staying below NA_3. We also select the first lens among available lenses to make NA_1 as large as possible while staying below NA_2. As the result, f_1, NA_1, f_2, NA_2 were determined as shown in Table 1.

	focal length	NA	diameter
first lens	8.0 mm	0.50	8.68 mm
second lens	3.1 mm	0.68	5.1 mm
APD sensitive area		0.78	0.175 mm

Table 1. Parameters of optical components

Taking into account of geometrical structure of lens mounts and fixing both of the first and second lens, the distance d_{12} should be longer than 3 mm. In this case, we determined $d_{12} = 5$ mm. The remaining parameters in this optical system were NA_0, ϕ_0, the distance d_1 between the fiber end and the first lens, and the distance d_2 between the second lens and the APD. These parameters can be determined in two steps as follows. In the first step, NA_0 and ϕ_0 are optimized to maximize the FOM expressed as Eq. (2) under the conditions $f_1 = d_1$ and $f_2 = d_2$. In the second step, d_1 and d_2 are optimized to maximize $\epsilon(d_1, d_2, \phi_0, NA_0)$ for given values of NA_0 and ϕ_0 obtained in the first step.

3.3 Determination of parameters in the optical system

The optimization procedure requires specific values of $\eta(NA_0)$ and $\epsilon(NA_0, \phi_0)$ for obtaining the FOM. Since the values of $\eta(NA_0)$ were given from Eq. (1) as shown in Fig. 3, it is necessary to calculate the values of $\epsilon(NA_0, \phi_0)$ at $d_1 = f_1$ and $d_2 = f_2$. The simple method for obtaining the efficiency is a statistical simulation by ray tracing.

The optical fiber end can be considered as a light source with ϕ_0 and NA_0. To calculate the efficiency, an event of light emission is randomly generated at an arbitrary position within ϕ_0 and at an arbitrary direction within NA_0. Subsequently, the final state of light at the APD sensitive area is calculated by the transformation of the initial state based on ray tracing. We repeat a procedure containing the generation of one event in the light emission and the subsequent transformation to the final state at the APD via intermediate states at the first lens and the second lens, and count the number of the events, where the conditions at the first lens NA_1, ϕ_1, at the second lens NA_2, ϕ_2, and at the APD NA_3 and ϕ_3 are all fulfilled. The efficiency can be obtained as the ratio of number of counts to the total number of event generation.

The propagation of light can be described with matrix formalism in paraxial optics (Yariv, 1997). The light at the initial state (r_i, r_i'), where r_i is a distance from an optical axis of optics and r_i' is a slope of the light direction, can be transferred by the following matrices, for example,

$$M_{free} = \begin{bmatrix} 1 & d \\ 0 & 1 \end{bmatrix}$$

$$M_{lens} = \begin{bmatrix} 1 & 0 \\ -\frac{1}{f} & 1 \end{bmatrix},$$

where M_{free} is the transfer matrix of free-space propagation far from the distance d and M_{lens} is the transfer matrix of thin lens with the focal length f. Any state of light expressed as the vector form (r, r') can be transferred by any combination of the transfer matrices. Therefore, the transfer matrix $M_{coupling}$ describing the optical system shown in Fig. 4 can be expressed as follows,

$$M_{coupling} = \begin{bmatrix} 1 & d_2 \\ 0 & 1 \end{bmatrix} \cdot \begin{bmatrix} 1 & 0 \\ -\frac{1}{f_2} & 1 \end{bmatrix} \cdot \begin{bmatrix} 1 & d_{12} \\ 0 & 1 \end{bmatrix} \cdot \begin{bmatrix} 1 & 0 \\ -\frac{1}{f_1} & 1 \end{bmatrix} \cdot \begin{bmatrix} 1 & d_1 \\ 0 & 1 \end{bmatrix}$$

$M_{coupling}$ can transfer the initial state (r_i, r_i') at the optical fiber end to the final state (r_f, r_f') at the APD sensitive area.

Thus, the values of ϵ were obtained by the statistical method with the transfer matrix $M_{coupling}$, where random numbers for setting initial states were generated with the software package based on algorithm of Mersenne Twister (Matsumoto & Nishimura, 1998). The calculated results are shown as a function of NA_0 in Fig. 5.

Fig. 5 shows that ϵ is decreasing with increasing NA_0 and ϕ_0. As a consequence, ϵ becomes 100 % at $\phi_0 = 0.4$ mm and $NA_0 \leq 0.25$. However, the reduction of ϕ_0 and NA_0 makes both η

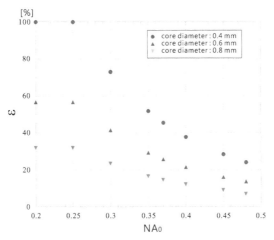

Fig. 5. Calculated values of coupling efficiency ϵ as a function of NA_0. The solid circles indicate the values at $\phi_0 = 0.4$ mm, the solid triangles are the values at $\phi_0 = 0.6$ mm, and the solid inverse triangles are the values at $\phi_0 = 0.8$ mm.

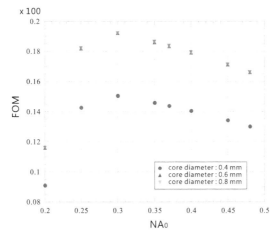

Fig. 6. Calculated values of FOM as a fuction of NA_0. The solid circles indicate the values at $\phi_0 = 0.4$ mm, the solid triangles are the values at $\phi_0 = 0.6$ mm, and the solid inverse triangles are the values at $\phi_0 = 0.8$ mm.

and the number of emitted photons smaller. Therefore, the calculation of the FOM is necessary for the optimization of NA_0 and ϕ_0.

The plot of the FOM as a function of NA_0 is shown in Fig. 6. One can easliy see that the FOM is maximal at $NA_0 = 3.0$. In addtion, the value of the FOM for $\phi_0 = 0.6$ mm is almost same as the one for $\phi_0 = 0.8$ mm. This means that the FOM is saturated for larger diameters than $\phi_0 = 0.6$ mm because the size of the transferred image at the APD is larger than the APD sensitive area.

Optical fibers with $NA_0 = 3.0$ are not easily obtainable, whereas optical fibers with $NA_0 = 2.5$ and $NA_0 = 3.7$ are readily available. Fig. 6 shows that the slope above $NA_0 = 3.0$ is flatter than the one below. In the upper part, the misalignment or imperfection of the optical system has less influence on the coupling efficiency. Therefore, we selected $\phi_0 = 0.6$ mm and $NA_0 = 3.7$.

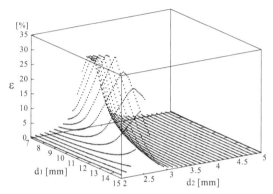

Fig. 7. Calculated values of ϵ as fuctions of d_1 and d_2 for given values of $NA_0 = 0.37$ and $\phi_0 = 0.6$ [mm].

In a second step, d_1 and d_2 were optimized to maximize $\epsilon(d_1, d_2)$ for $NA_0 = 0.37$ and $\phi_0 = 0.6$ mm. Fig. 7 shows the calculated values of ϵ as functions of d_1 and d_2. From the peak of ϵ in Fig. 7, the results of $d_1 = 11.6$ mm and $d_2 = 2.7$ mm were obtained for $NA_0 = 0.37$ and $\phi_0 = 0.6$. These parameters provide the maximum $\epsilon(d_1, d_2)$ of 33.33 %, $\eta(NA_0)$ of 1.95 % and FOM $\times 100$ is 0.234 which value is higher than the maximum in Fig. 6.

4. Application to the ATP sensing system

4.1 Luciferin-luciferase reaction

Bioluminescence in living organisms, such as fireflies and some marine bacteria, typically occurs due to the optical transition from the excited state to the ground state of oxidized luciferin molecules produced by the luciferin-luciferase reaction under the catalytic activity of luciferase molecules. This reaction can be expressed by the following sequence of reaction steps:

$$E + LH_2 + S \rightleftharpoons C + PP_i$$
$$C + O_2 \rightarrow E + P^* + AMP + CO_2$$
$$P^* \rightarrow P + \gamma$$

where E indicates luciferase, LH_2 luciferin, PP_i pyrophoric acid, C is an enzyme-substrate compound E · LH_2-AMP, AMP adenosine monophosphate, P oxidized luciferin, and γ a photon (DeLuca, 1976). The emission of one photon at the position of luciferase molecule indicates the use of the energy of one ATP molecule.

The immobilization of luciferase molecules at the optical fiber end enables us to sense the presence of ATP around the fiber end using single photon detection. For this purpose, we used a compound protein containing a silica-binding protein (SBP) molecule and a luciferase molecule (SBP-luciferase), which were recently synthesized by Taniguchi and co-workers (Taniguchi et al., 2007). This protein makes it possible to immobilize a luciferase molecule on the optical fiber end via a SBP molecule while retaining its activity. The spectrum of the emitted photons shows a central wavelength of 550 nm and a width of about 100 nm (Denburg et al., 1969), (Ugarova & Brovko, 2002). Since the APD module has the large quantum efficiency for the photons from the luciferin-luciferase reaction, the APD photon counting module is suitable for ATP sensing.

4.2 Reaction-diffusion differential equation

Fig. 8 shows the enzyme reaction which describes the sequence of reactions.

$$s + e \underset{k_-}{\overset{k_+}{\rightleftharpoons}} c \overset{h_1}{\longrightarrow} p + e + n_\gamma$$

s : ATP (S)
e : luciferaze (E)
p : oxyluciferin (P)
n_γ : photon (γ)
c : E · LH$_2$-AMP
k_+, k_- : kinetic constants
h_1 : reaction constant

Fig. 8. Enzyme reaction describing luciferin-luciferase reaction

Here, s, e, c, p, n_γ in Fig. 8 represents a concentration of ATP, luciferase, E · LH$_2$-AMP, oxidized luciferin, and emitted photon, respectively. In addition, k_+, k_- are kinetic constants for equilibrium and h_1 is a reaction constant.

In a solution containing nonlocalized homogeneously dispersed luciferase and ATP, the Michaelis-Menten theory can be simply applied to the above reaction. In the presence of enough luciferin molecules in the solution, a rate of emitted photons at steady state v_γ can be expressed as the Michaelis-Menten formula,

$$v_\gamma = \frac{V_0 S}{S + K_M}, \tag{3}$$

where V_0 is a maximum reaction rate which is equivalent to a product of the number of luciferase molecule and h_1, K_M is the Michaels constant expressed as $K_M = \frac{k_- + h_1}{k_+}$ and S is the ATP concentration.

In the fiber-based system for sensing dispersed ATP molecules, on the other hands, an ATP-flow generated by a gradient of ATP concentration around the luciferase-terminated fiber end carries ATP molecules to the vicinity of immobilized SBP-luciferase molecules. The ATP molecule is bound to the immobilized SBP-luciferase molecule near them and subsequently contributes the luciferin-luciferase reaction at this fiber end. To calculate the rate of emitted photons, therefore, it is necessary to consider not only a reaction rate but also an ATP diffusion rate.

The series of reaction-diffusion equations describing the enzyme reaction shown in Fig. 8 can be expressed as,

$$\frac{ds}{dt} = D \left(\frac{\partial^2 s}{\partial x^2} + \frac{\partial^2 s}{\partial y^2} \right) + R(s, e, c)$$

$$\frac{de}{dt} = -k_+ se + k_- c + h_1 c$$

$$\frac{dc}{dt} = k_+ se - k_- c - h_1 c$$

$$\frac{dn_\gamma}{dt} = h_1 c, \tag{4}$$

where D is a diffusion constant of ATP in water and $R(s, e, c)$ is expressed as follows.

$$R(s, e, c) \equiv \begin{cases} -k_+ se + k_- c & (x, y \in \Gamma_{fiber}) \\ 0 & (\text{otherwise}) \end{cases}$$

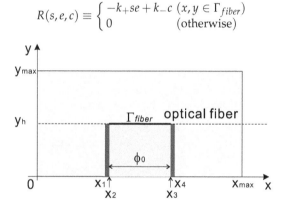

Fig. 9. Coordinate system of x and y around the fiber end.

Fig. 9 shows the definition of the coordinate system of x and y . Here, it should be noted that $s(x, y, t)$ is a function of x, y, t and $e(t)$ and $c(t)$ are functions of only t. Γ_{fiber} is defined as an area where the reaction occurs and equivalent to the core part ϕ_0 in the optical fiber. In this coordinate system, it can be represented as the interval of $x_2 \leq x \leq x_3$ with $y = y_h$. The both intervals of $x_1 \leq x \leq x_2$ and $x_3 \leq x \leq x_4$ with $y = y_h$ represent the parts of clad in the optical fiber.

By dividing the space around the fiber end into finite spatial steps and also dividing the time into finite time steps, we can numerically solve the series of reaction-diffusion equation Eq. (4) under the boundary conditions presented in Table 2. Here, s_0 in Table 2 is an initial ATP concentration that should be uniform into the whole solution before starting the reaction. The numerical solutions can be given as time evolution of spatial distribution of concentration $s(x_i, y_i, t_i)$, $e(t_i)$, and $c(t_i)$ and the rate of emitted photons can be obtained from $h_1 c(t_i)$. The peak of the photon-emission rate in time corresponds to the reaction rate given by the Michaelis-Menten formula.

In order to check the possibility that the emission rates are limited by the diffusion rate of ATP, we numerically obtained the peak values of the photon-emission rate at various ATP concentration and compared to the reaction rate calculated by Michaelis-Menten formula. To obtain numerical solutions at each time step, spatial segmented equations derived from Eq.

region	$x = 0$	$0 \le x < x_1$ $y = 0$	$x = x_1$ $0 \le y < y_h$	$x_1 \le x < x_2$ $y = y_h$	$x_2 \le x \le x_3$ $y = y_h$
condition	$s = s_0$	$s = s_0$	$\left.\dfrac{ds}{dx}\right|_{x=x_1} = 0$	$\left.\dfrac{ds}{dy}\right|_{y=y_h} = 0$	$\left.\dfrac{ds}{dy}\right|_{y=y_h} = 0$
region	$x_3 < x \le x_4$ $y = y_h$	$x = x_4$ $0 \le y < y_h$	$x_4 < x \le x_{max}$ $y = 0$	$x = x_{max}$	$y = y_{max}$
condition	$\left.\dfrac{ds}{dy}\right|_{y=y_h} = 0$	$\left.\dfrac{ds}{dx}\right|_{x=x_4} = 0$	$s = s_0$	$s = s_0$	$s = s_0$

Table 2. boundary conditions for solving Eq. (4). s_0 is the initial ATP concentration.

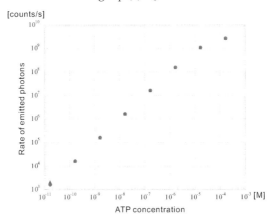

Fig. 10. Comparison of the results calculated from numerical solutions of reaction-diffusion equations to the results of Michaelis-Menten formula. Closed triangles indicate values of ATP diffusion process using typical values of kinetic constants, closed circles are values given by Michaelis-Menten formula.

(4) were solved in terms of time using the software package of ordinary differential equations DASKR (Brown et al., 1998), where the ATP diffusion constant was $D = 0.5 \times 10^{-5}$ cm^2/s (Aflao & DeLuca, 1987), the geometrical parameters were $x_{max} = 1.4$ mm, $y_{max} = 2$ mm, $y_h = 1$ mm, $x_1 = 0.38$ mm, $x_4 = 1.02$ mm, the kinetic constants were $k_+ = 20000$ M^{-1}s^{-1}, which was estimated from the typical buildup time of 0.3 s (DeLuca & McElory, 1974), and $k_- = 0.515$ s^{-1}, which was calculated with the relational form $K_M \cdot k_+ - h_1$. As other parameters, we used $h_1 = 0.125$ s^{-1} (Branchini et al., 2001), $K_M = 3.2 \times 10^{-5}$ M, and the surface density of luciferase molecule $v_0 = 9.03 \times 10^{10}$ mm^{-2} (Taniguchi et al., 2007), which were also used for Michaelis-Menten formula. The Michaelis constants $K_M = 3.2 \times 10^{-5}$ M was obtained from data analysis of counts of detected photons, which describes in Sec. 4.

Fig. 10 shows the results of two kinds of calculations. The closed triangles indicate the values obtained from numerical solutions of Eq. (4), whereas the closed circles are the values deduced from Michaelis-Menten formula. The results of ATP diffusion process have good agreements with ones given by the Michaelis-Menten theory. Therefore, we can conclude that the ATP diffusion is not a rate-limiting process for the present rate of the chemical reaction.

4.3 Measurement of the sensitivity

4.3.1 Immobilization of luciferase

Before immobilizing the luciferase molecules, we cut optical fiber and cleaned the cut surface with ethanol and Tris buffer (0.25 mM Tris-HCl with 0.15 M NaCl). After cleaning, the surface was immersed in a solution of SBP-luciferase and was left at a temperature of 3°C to 6°C for a period of about two hours.

4.3.2 Sample solutions

The samples were a 1:4:4:31 mixture of 20 mM D-luciferin solution, Tris buffer solution(250 mM Tris-HCl mixed with 50 mM $MgCl_2$), ATP solution, and distilled water. Several solutions of ATP with different ATP concentrations were made by diluting the ATP standard in ATP Bioluminescence Assay Kit CLS II manufactured by Roche Co. Ltd. Thus, a series of sample solutions with different ATP concentrations were prepared in advance. An additional sample without ATP was also produced by mixing distilled water instead of the ATP solution. This sample was measured in order to obtain a background before the ATP measurements.

4.3.3 Data taking system

TTL pulses outputted from the APD photon counting module were counted by a PC card installed in a personal computer(PC). The number of pulses occurring during 10 s were recorded every 10 s by the PC

4.3.4 Results

The time dependence of photon counts per 10-s interval were measured after the luciferase-terminated end were immersed in the sample solutions with various ATP concentration. The results for 100 $\mu\ell$ solution from 1.65×10^{-4} M to 1.65×10^{-9} are shown in Fig. 11.

The photon counts increase and reach a maximum at about 150 s after immersion. Then, they decrease very slowly to the background level. Therefore, for obtaining high sensitivity, it is practical to continue the measurement for about 300 s after the luciferase-terminated end is immersed in the solution. A background level of approximately 120 s^{-1} was determined as an average of counts for background data. It was found to correspond well to the dark counts of the photon counting module.

ATP concentration	integrated counts	BG-subtracted counts	statistical errors
no ATP (background)	34490		
1.65×10^{-11} M	34664	174	262.97
1.65×10^{-10} M	35346	856	264.27
1.65×10^{-9} M	46770	12280	285.06
1.65×10^{-8} M	139300	104810	416.88

Table 3. Integrated counts for ATP sample solutions

Table 3 presents integrated counts of detected photons during 300 s from the origin of time. Statistical errors are estimated as one standard deviation assuming Poisson distribution. From Table 3, the sensitivity in this optical fiber-based system is limited to 1.65×10^{-10} M, which

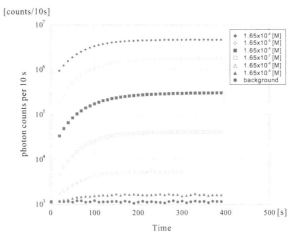

Fig. 11. Measured photon counts per 10 s as a function of time at ATP concentration 1.65×10^{-4} M, 1.65×10^{-5} M, 1.65×10^{-6} M, 1.65×10^{-7} M, 1.65×10^{-8} M, 1.65×10^{-9} M, which are represented by closed diamonds, open diamonds, closed squares, open squares, open triangles, closed triangles, respectively. Closed circles indicate background counts.

corresponds to a number of ATP molecules of about 10^{-14} mol in the 100 $\mu\ell$ solution. We also ascertained that our system is sensitive to the 10^{-10} M level even with a $10\mu\ell$ solution, in which the absolute ATP concentration is 10^{-15} mol.

In order to check the ATP concentration dependence of the photon counting rate at maximum, the average of counts in six 10-s intervals around the time at which the counting rate became maximal was calculated for each ATP concentration. The results are indicated by solid circles in Fig. 12. This figure shows that the lower limit of the sensitivity is 1.65×10^{-9} M taking into account of statistical errors. If the statistical errors reduce to half of the present ones, which means that statistics at each point becomes four times as large as 60-s interval, the detection of 1.65×10^{-10} M also becomes feasible. This is consistent with the results presented in Table 3. The analysis of fitting data points in Fig. 12 provided the Michaelis constant of 3.2×10^{-5} M.

4.3.5 Discussion

In Fig. 12, the open squares show predictions of the counting rate estimated from the results of numerical solutions shown in Fig. 10 with the total detection efficiency in our detection system, which can be expressed as $\epsilon_{total} = \eta \cdot \epsilon \cdot \eta_{qe}$. Its value was calculated to be 0.00354 for photons with the wavelength 550 nm. Comparing the experimental data with the predictions, there are disagreements amounting to roughly one order of magnitude.

The results suggest the ATP diffusion does not limit the detection limit. Therefore, the possible reasons for the discrepancies may be the following.

(1) The rate of the chemical reaction per luciferase molecule itself becomes small.
(2) The number of immobilized active molecules is lower than expected.

We have also confirmed that the buildup time in Fig. 11 is about 30 times longer than that in a solution containing only nonlocalized homogeneously dispersed SBP-luciferase

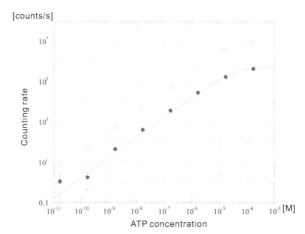

Fig. 12. Measured photon counting rate as a function of ATP concentration. Solid line is a curve obtained by fitting data with the Michaelis-Menten formula. Open squares are values estimated from the results of numerical solutions shown in Fig. 10 with total detection efficiency $\epsilon_{total} = 0.00354$.

molecules. There are also other experimental results claiming the decrease in acitivity of immobilized luciferase molecues (Nishiyama et al., 2008), (Tanaka et al., 2011). The reason of this discrepancy is still an open question. These results suggest that it will be necessary to study the reaction at the end of the optical fiber in more detail in order to improve the sensitivity.

5. Summary

We introduced a method of high-sensitivity detection of bioluminescence at an optical fiber end for an ATP sensor as an efficient alternative to direct detection of bioluminescence from a sample solution. For efficiently detection of the bioluminescence, we have constructed an optical fiber-based system, where the luciferase molecules are immobilized on the optical fiber end and the other end is optically coupled to an APD-type photon counting module. We have evaluated the sensitivity of the ATP detection of this system and found that the optimization of the optical coupling system and the use of the SBP-luciferase molecules provide the detection limit of 10^{-10} M, which allows us to detect the absolute ATP concentration of 10^{-15} mol with a $10\mu\ell$ solution. To improve the sensitivity, it is necessary to study the details of the reaction at luciferase-terminated end of the optical fiber.

6. Acknowledgements

We are grateful to Dr. Kazutaka Nomura for providing the SBP-luciferase molecules and Prof. Holger F. Hofmann for reading the manuscript. We also express our thanks to Dr. Tetsuya Sato, Dr. Koichiro Sadakane, and Dr. Shuhei Miyoshi for their help in setting up the experiments. This work has been partially supported by the International Project Center for Integration Research on Quantum, Information, and Life Science of Hiroshima University and the Grant-in-Aid for Scientific Research (C)(19560046) of Japanese Society for the Promotion of Science, JSPS.

7. References

Andreotti P. E. & Berthold F. (1999). Application of a new high sensitivity luminometer for industrial microbilogy and molecular biology, *Luminescence*, Vol. 14: 19-22, ISSN 15227235

Arnold M. A. (1991). Fluorophore- and Chromophore-Based Fiberoptic Biosensors, In: *Biosensor principles and applications*, Blum L. J. & Coulet P. R. (Ed.), 195-211, Marcel Dekker, 1991, ISBN 0824785460, New York

Aflao C. & DeLuca M. (1987). Continuous Monitoring of Adenosine 5'-Triphosphate in the Microenvironment of Immobilized Enzymes by Firefly Luciferase, *Biochemistry*, Vol. 26: 3913-3920, ISSN 00062960

Blum L. J.; Coulet P. R.; Gautheron D. C. (1984). Collagen Strip with Immobilized Luciferase for ATP Bioluminescent Determination, *Biotechnol. Bioeng.*, Vol. 27: 232-237, ISSN 00063592

Blum L. J.; Gautier S. M. (1991). Bioluminescence- and Chemiluminescence-Based Fiberoptic Sensors, In: *Biosensor principles and applications*, Blum L. J. & Coulet P. R. (editors), 213-247, Marcel Dekker, 1991, ISBN 0824785460, New York

Branchini B. R.; Magyar R. A.; Murtiashaw M. H.; Portier N. C. (2001). The Role of Active Site Residue Arginie 218 in Firefly Luciferase Bioluminesence, *Biochemistry*, Vol. 40: 921-925, ISSN 00062960

Brown P. N.; Hindmarsh A. C.; Petzold L. R. (1998). Consistent initial condition calculation for differential-algebraic systems, *SIAM J. Sci. Comput.*, Vol. 19: 1495-1512, ISSN 10648275. URL: http://www.netlib.org/ode/

DeLuca M. & McElory W. D. (1974). Kinetics of the Firefly Luciferase Catalyzed Reactions, *Biochemistry*, Vol. 13: 921-925, ISSN 00062960

DeLuca M. (1974). Firefly luciferase, In: *Advances in Enzymology*, Meister A. (Ed.), Vol. 44: 37-68, Wiley, 1976, ISBN 9780471591795, New York

Denburg J. L.; Lee R. T.; McElory W. D. (1969). Substrate-Binding Properties of Firefly Luciferase 1. Luciferin-binding Site, *Arch. biochem. biophys.*, Vol. 134: 381-394, ISSN 00039861

Fraga H. (2008). Firefly luminescence: A histrical perspective and recent developments, *Photochem. Photobio. Sci.*, Vol. 7: 146-158, ISSN 1474905

Gourine A. V.; Laudet E.; Dale N.; Spyer K. M. (2005). ATP is a mediator of chemosensory transduction in the central nervous system, *Nature*, Vol. 436: 108-111, ISSN 00280836

Iinuma M.; Ushio Y.; Kuroda A.; Kadoya Y. (2009). High Sensitivity Detection of ATP Using Bioluminescence at An Optical Fiber End, *Electronics and Communications in Japan*, Vol. 92: 53 59, ISSN 03854221

Konno T.; Ito T.; Takai M.; Ishihara K. (2006). Enzymatic photochemical sensing using luciferase-immobilized polymer nanoparticles covered with artificial cell membrane, *J. Biometer. Sci. Polymer. Edn.*, Vol. 17: 1347-1357, ISSN 09205063

Lee Y.; Jablonski I.; DeLuca M. (1977). Immobilization of Firefly Luciferase on Glass Rods: Properties of the Immobilized Enzyme, *Anal. Biochem.*, Vol. 80: 496-501, ISSN 00032697

Matsumoto M. & Nishimura T. (1998). Mersenne Twister: A 623-dimensionally equidistributed uniform pseudorandom number generator, *ACM Trans. model. comput. simul.*, Vol. 8: 3-30, ISSN 10493301. URL: http://www.math.sci.hiroshima-u.ac.jp/ m-mat/MT/emt.html

Nakamura M.; Mie M.; Funabashi H.; Yamamoto K.; Ando J; Kobatake E. (2006). Cell-surface-localized ATP detection with immobilized firefly luciferase, *Anal. Biochem.*, Vol. 352: 61-67, ISSN 00032697

Nishiyama K.; Watanabe T.; Hoshina T.; Ohdomari I. (2008). The main factor of the decrease in activity of luciferase on the Si surface, *Chem. Phys. Lett.*, Vol. 453: 279-282, ISSN 00092614

Ribeiro A. R.; Santos R. M.; Rosário L. M.; Gil M. H. (1998). Immobilization of Luciferase from a Firefly Lantern Extract on Glass Strips as an Alternative Strategy for Luminescent Detection of ATP, *J. Biolumin. Chemilumin.*, Vol. 13: 371-378, ISSN 08843996

Stanley P. E. (1992). A survey of more than 90 Commercially Available Luminometers and Imaging Devices for Low-light Measurements of Chemiluminescence and Bioluminescence, Including Instruments for Manual, Automatic and Specialized Operation, for HPLC, LC, GLC and Microtitre Plates, *J. Biolumin. Chemilumin.*, Vol. 7: 77-108, ISSN 08843996

Tanaka. R.; Takahama E.; Iinuma M.; Ikeda T.; Kadoya Y.; Kuroda A. (2011). Bioluminescent Reaction by Immobilized Luciferase, *IEEJ Trans. EIS*, Vol. 131: 23-28, ISSN 03854221

Taniguchi K.; Nomura K.; Hata Y.; Nishimura Y.; Asami Y.; Kuroda A. (2007). The Si-tag for immobilizing proteins on a silica surface, *Biotechnol. Bioeng.*, Vol. 96: 1023-1029, ISSN 00063592

Tanii T.; Goto T.; Iida T.; Koh-Masahara M.; Ohdomari I. (2001). Fabrication of Adenosine Triphosphate-Molecule Recognition Chip by Means of Bioluminous Enzyme Luciferase, *Jpn. J. Appl. Phys.*, Vol. 40: L1135-L1137, ISSN 00214922

Tsuboi Y.; Furuhata Y.; Kitamura N. (2007). A sensor for adenosine triphosphate fabricated by laser-induced forward transfer of luciferase onto a poly(dimethylsiloxane) microchip, *Appl. surf. sci*, Vol. 253: 8422-8427, ISSN 01694332

Ugarova N. N.; Brovko L. Y. (2002). Protein structure and bioluminescent spectra for firefly bioluminescence, *Luminescence*, Vol. 17: 321-330, ISSN 15227235

Worsfold P. J.; Nabi A. (1986). Bioluminescent assays with immobilized firefly luciferase based on flow injection analysis, *Anal. Chim. acta*, Vol. 179: 307-313, ISSN 00032670

Yariv A. (1997). *Optical Electronics in Modern Commnications Fifth Edition*, Oxford University Press, ISBN 0-19-510626-1, New York.

Yotter R. A.; Lee L. A.; Wilson D. M. (2004). Sensor Technologies for Monitoring Metabolic Activity in Single Cells - Part I: Optical Methods, *IEEE Sensors Journal*, Vol. 4: 395-411, ISSN 00032670

9

Fiber Optics for Thermometry in Hyperthermia Therapy

Mario Francisco Jesús Cepeda Rubio,
Arturo Vera Hernández and Lorenzo Leija Salas
Centro de Investigación y de Estudios Avanzados del Instituto Politécnico Nacional,
Department of Electrical Engineering/Bioelectronics
Mexico

1. Introduction

A review of electromagnetic hyperthermia ablative therapies, a temperature measurements during microwave ablation using optical fiber temperature sensors and a fiber optic temperature sensor developed for a hyperthermia laboratory are summarized in this chapter. Hyperthermia also called thermotherapy or thermal therapy is a type of cancer treatment in which body tissue is exposed to high temperatures. All methods of hyperthermia include transfer of heat into the body from an external energy source and are currently under study, including local, regional, and whole-body hyperthermia. The application of heat to treat patients with a malignant tumor is not a novel concept. The Edwin Smith papyrus explains the topical application of heated metallic implements or hot oil that were used approximately 5000 years ago to treat patients with tumors (Izzo et al., 2001). Local hyperthermia is used to heat a small area. It involves creating very high temperatures that destroy the cells that are heated. Research has shown that high temperatures can damage and kill cancer cells, usually with minimal injury to normal tissues (van der Zee, 2002). By killing cancer cells and damaging proteins and structures within cells, hyperthermia may shrink tumors (Hildebrandt et al., 2002). Numerous clinical trials have studied hyperthermia in combination with radiation therapy and/or chemotherapy. Many of these studies, have shown a significant reduction in tumor size when hyperthermia is combined with other treatments (Wust et al., 2002). Otherwise, ablation or high temperature hyperthermia, including lasers and the use of radiofrequency, microwaves, and high-intensity focused ultrasound, are gaining attention as an alternative to conventional surgical therapies (van Esser et al., 2007). Each of these techniques works differently, the objective is to heat tissue to a temperature 50°C above to destroy cells within a localized section of a malignant tumor. Frequently, temperature monitoring during thermal therapy has been achieved using interstitial thermocouples, thermistors, or fiber optics probes. Fiber optics technology should exploit one or more of electromagnetic immunity, noninvasiveness, chemical immunity, small size, or the capacity for distributed measurement. Fiber optic thermometers are used when electrical insulation and EMI immunity are necessary. The most relevant application involves tissue-heating control during microwave ablation (MWA) or radiofrequency ablation (RFA) hyperthermia therapy for cancer treatment (Mignani & Baldini, 1996).

1.1 Radiofrequency ablation

RFA works by converting radiofrequency waves into heat through ionic vibration. Alternating current passing from an electrode into the surrounding tissue causes ions to vibrate in an attempt to follow the change in the direction of the rapidly alternating current. It is the ionic friction that generates the heat within the tissue and not the electrode itself. The higher the current, the more vigorous the motion of the ions and the higher the temperature reached over a certain time, eventually leading to coagulation necrosis and cell death. The ability to efficiently and predictably create an ablation is based on the energy balance between the heat conduction of localized radiofrequency energy and the heat convection from the circulation of blood, lymph, or extra and intracellular fluid. The amount of radiofrequency produced heat, is directly related to the current density dropping precipitously away from the electrodes, thus resulting in lower periphery temperatures. The goal of radiofrequency ablation is to achieve local temperatures that are lethal to the targeted tissue. Generally, thermal damage to cells begins at 42°C; and once above 60°C, intracellular proteins are denatured, the lipid bilayer melts, and irreversible cell death occurs (Simon et al., 2005).

1.2 Microwave ablation

Water molecules are polar; the electric charges on the molecules are not symmetric. The orientation and the charges on the atoms are such that the oxygen side has a negative charge, and the hydrogen side of the molecule has a positive charge. When an oscillating electric charge from microwave radiation interacts with a water molecule, it causes the molecule to flip. Microwave radiation is specially tuned to the natural frequency of water molecules to maximize this interaction. As a result of the radiation hitting the molecules, the electrical charge on the water molecule flips back and forth 2–5 billion times a second depending on the frequency of the microwave energy. Temperature is a measure of how fast molecules move in a substance, and the vigorous movement of water molecules raises the temperature of water. Thus, electromagnetic microwaves heat matter by agitating water molecules in the surrounding tissue, producing friction and heat, thus inducing cellular death via coagulation necrosis (Goldberg & Gazelle, 2001).

2. Fiber optics thermometer

One of the first commercial fiber optic temperature probe was a fluorescence-based temperature sensor introduced in the early 1980s by the Luxtron Corporation of Mountain View, CA. Successors to these sensors are commercially available and are effective, approach to solving measurement problems (Culshaw, 1999). These include monitoring temperature patterns, examining temperature distribution in power transformer oils, and similar areas where the issue is the operation of a reasonably precise temperature probe within very high electromagnetic fields. In such circumstances, a metallic temperature sensor either distorts the electromagnetic field significantly or is subjected to very high levels of interference, producing spurious evaluations. Other applications sectors exploit the small size or chemical passivity of the device, including operation within corrosive solvents or examination of extremely localized phenomena such as laser heating or in determining the selectivity of radiation and hyperthermia treatments.

2.1 Fiber optics for thermometry in hyperthermia

Temperature probes that employ optical fibers were introduced for clinical hyperthermia treatment in the 1980s. Clini-therm Corporation, Dallas, TX founded in 1978 (no longer in business), at that time the company was working to develop and manufacture a system for hyperthermia therapy. The system incorporates fiber optic temperature probes which are inserted into a malignant tumor and surrounding healthy tissue through catheters. The probes detect changes in temperature during hyperthermia treatment.

(Vaguine et al., 1984) developed another probe. It uses a small crystal of the semiconductor gallium arsenide (GaAs) whose optical absorption at a specifically chosen wavelength is sensitively related to the crystal's temperature. The amount of optical signal returned after passage through the sensor may be detected and electronically translated, after calibration, into an indication of the probe's temperature. The system has up to 12 temperature sensors which are packaged in two basic probe configurations: a single-sensor probe with a length of 1.2 m and a diameter of 0.6 mm; and a four-sensor linear array probe with a length of 1.2 m, diameter of 1.1 mm, and spacing of 1.5 cm between adjacent sensors.

(Katzir et al., 1989) reported a fiber optic radiometer system, which is based on a nonmetallic, infrared fiber probe, which can operate either in contact or noncontact mode. In preliminary investigations, the radiometer worked in a strong microwave or radiofrequency field, with an accuracy of ± 0.5°C. The fiber optic thermometer was used to control the surface temperature of objects within ± 2°C.

(Wook et al., 2009) reported a measured temperature distribution using the infrared optical fibers during radiofrequency ablation. Infrared radiations generated from the water around inserted single-shaft plain electrode are transferred by the three silver halide optical fibers and are measured by a thermopile sensor array. Also, the output voltages of the thermopile sensor array are compared with those of the thermocouple recorder. It was concluded that is expected that a non-contact temperature sensor using the infrared optical fibers can be developed for the real time temperature monitoring during RFA treatments based on the results of this study.

Intelligent Optical Systems, Inc., has reported in 2009 an Optical Fiber Temperature (OFT) probe that will provide 10 points of high-resolution temperature data in a single fiber. The OFT probe is an alternative to thick bundles of arrayed single point temperature. According to the company, this new probe will provide a realistic, cost effective, and efficient technique for distributed temperature monitoring and profiling in treated tissues, and use of the OFT probe can reduce complications as a result of excessive heat exposure and damage to the healthy tissue surrounding a tumor. As microwave hyperthermia becomes increasingly popular for the treatment of benign prostatic hyperplasia and other benign or malignant tumors. The OFT provides multipoint and self-calibrating temperature sensing. Measures tumor temperatures at multiple points along a 5 to 10 cm sensor length, with the ability to pinpoint target areas within 0.5 cm. Monitors these temperatures over a large range, including 35°C to 55°C range, with a 0.1°C temperature resolution. Readily interfaces with clinical hyperthermia catheters because of its size (0.25mm thick) and thus will be adaptable to a variety of clinical conditions.

(Saxena & Hui, 2010) reported a polymer-coated fiber Bragg grating (PFBG) technology that provides a number of FBG thermometry locations along the length of a single optical fiber.

The PFBG probe was tested in an environment designed to approximate the microwave exposure that might be encountered during clinical hyperthermia treatments. According to the authors, this offers an enabling alternative to either scanning or bundled single point temperature probes for distributed thermometry in clinical applications.

3. Fiber optics thermometry system for a hyperthermia laboratory

In this section, we present a fiber optic temperature sensor developed for a hyperthermia laboratory. Many benefits coming from hyperthermia depends on the ability to induce and to measure high temperatures locally. The heating is carried out via local irradiation of energy, mainly by means of ultrasonic or electromagnetic waves. To be able to manage the irradiation parameters it is necessary to know the temperature to which the radiated tissue rises, that makes practically indispensable to have a highly reliable system of temperature measurement. The measurement of temperatures is carried out with thermometry systems that can be invasive, or non invasive. The use of the conventional sensors (thermistors, thermocouples, etc.) it is not satisfactory in some applications, just as it is the case of the therapies for hyperthermia with microwaves. This is due to that currents and voltages are induced by electromagnetic interference in the metallic elements and a self-heating by induction appears. Both factors produce erroneous readings as a result in the measurements when using these sensors. When these conditions are presented it is necessary to build sensors denominated, as non-disturbing electromagnetic field, like those based on optical fibers. Among them we find fluorescent sensors, interferometric sensors, those based on the variation of intensity caused by absorption or reflection and the evanescent sensors. At the present time, there are in the world diverse laboratories dedicated to the experimental investigation in hyperthermia, where the electromagnetic radiation is used inside a controlled and safety ambient. In the Section of Bioelectronics of the CINVESTAV-IPN there is an automated laboratory (Chong et al., 2000), in which controlled radiation experiments with microwaves are carried out in the interval from 900 MHz to 8 GHz and pulsating wave ultrasound at 1 MHz. This radiation has an impact on a biological tissue substitute material, where heating is obtained by the interaction of radiation with the radiated material. The experiments are carried out inside an anechoic chamber to avoid the external electromagnetic interferences and at the liberation of the energy radiation to the environment. The thermometry system that is described is part of this automated laboratory (Pennisi et al., 2002). This system measures the temperature inside the phantom, which is heated through the electromagnetic and ultrasonic fields radiations that are being carried out.

3.1 Methodology

The system can be divided in four parts: the stage of temperature measurement using optical fiber sensors and the analog signal conditioning; the stage of automated positioning of the temperature sensors inside the phantom; the communication and control stage using two personal computers; and a general software for the visualization of the temperature distribution inside the phantom. Figure 1 shows a block diagram of the system.

3.1.1 Temperature sensor

This model is an approach based on the theory of the weakly guiding fibers, and don't take into account the length of the sensor. However, we can see that the experimental results

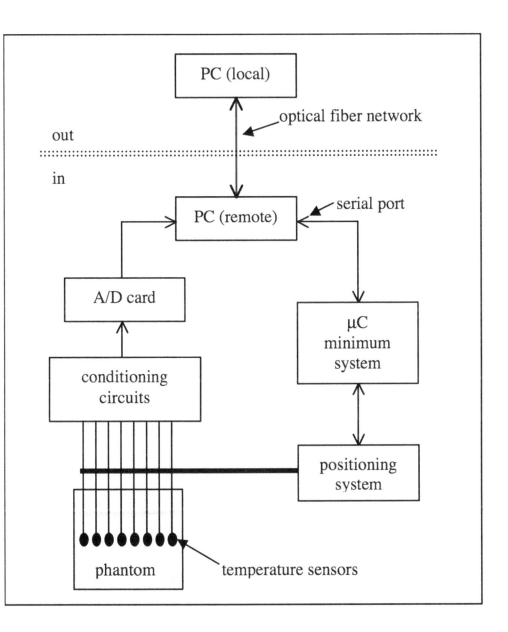

Fig. 1. Diagram of the hyperthermia automated laboratory. In the inner of the anechoic chamber, a PC controls the temperature sensors and the positioning system. A PC located outside the chamber for the visualization of the temperature distribution.

match with the response predicted by the model. Assuming that there is an allowable power flow in both core and cladding, the combined fiber power flow is:

$$P = P_1 + P_2 \text{ (Watts)} \tag{1}$$

where the subscripts 1 and 2 refer to the core and cladding respectively. The ratio of optical powers in core and cladding is:

$$\frac{P_1}{P_2} = \frac{3}{4} N_m^{1/2} - 1 \tag{2}$$

where N_m, represents the total number of free-space modes accepted and transmitted by a step-index fiber. The number of modes is related with the so-called v value by the relationship:

$$N_m = \frac{1}{2} v^2 \tag{3}$$

The v value, sometimes referred to the normalized frequency, is an important fundamental fiber property, because it contains the most important fiber parameters, and is defined by:

$$v = 2\pi \left(\frac{a_1}{\lambda} \right) \left(n_1^2 - n_2^2 \right)^{1/2} \tag{4}$$

where a_1 is the fiber core radius, λ is the free space wavelength of operation and n is the refractive index. The normalized power at the end of the fiber is then:

$$P_{norm} = 1 - \frac{P_2}{P} = 1 - \frac{2\sqrt{2}}{3\pi \left(\frac{a_1}{\lambda} \right) \left(n_1^2 - n_2(T)^2 \right)^{1/2}} \tag{5}$$

From the latter expression, we can see that the power transmitted along the fiber is dependent on the temperature, with an inverse square root relationship. Due to the refraction index wavelength dependence, we have evaluated three different wavelengths for the sensor operation (660 nm, 850 nm and 1300 nm), as well as several different types of oils, with a slightly different refraction index. For our range of interest the best results we have obtained were with vegetable oils, operating in the infrared range. The sensor was fabricated using a multimode patch-cord with ST connectors on both ends, which is commercially available. First, the patch-cord is divided in halves. A 30 mm portion of jacket and plastic buffers are mechanically removed from both ends. After that, a 20 mm glass tube with an external diameter of 1.1 mm is introduced through one of the ends, and both ends of the fiber are joined together with a fusion splicer (RXS-X74). Then the fiber without buffer of this portion is treated with an acid, to remove partially the glass cladding. The acid attack is stopped when the transmitted power falls 30% from its initial value. The glass tube is then located in the central part of the fiber and is filled with the oil. Finally, both ends of the glass tube are sealed with epoxy glue. Figure 2 shows a detailed outline of the probe described above.

The basic setup is showed on Figure 3. There is a lightwave multimeter (HP 8153A) configured with an optical source module (Hp8155IMM) and a power sensor module (HP

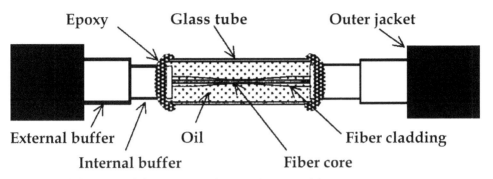

Fig. 2. Detailed outline of the constructed sensor (not in scale).

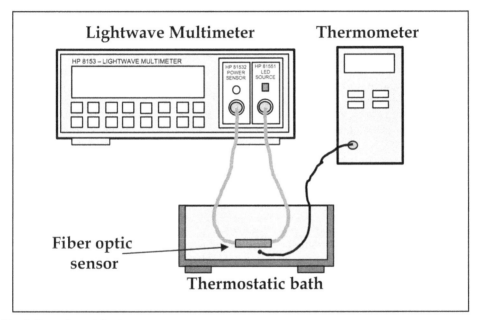

Fig. 3. Experimental setup.

81532A). The stability of the source was evaluated, and was +0.03 dB in a 6 h interval. The fiber optic sensor is connected between the source and the power sensor, and is immersed in a thermostatic bath. The wavelength of operation selected is 850 nm.

3.1.2 Positioning stage

This stage is constituted by a personal computer (PC), a microcontroller and a PC controlled automated positioning system. The PC is IBM compatible. It sends the data that correspond to a position selected by the user toward the minimum system. This PC is located inside the anechoic chamber and it is externally operated through an optical fiber communication system. The communication between the PC and the minimum system is carried out through the communications port COM1. The temperature sensors are displaced in an axis through the phantom by means of an automated positioning system. Two DC motors controlled by the minimum system by means of the feedback signal that provides an optical encoder impel this. The maximum lineal travel distance is of 35 cm, with a resolution of 2.5 mm. The physical construction of the positioning system is built on a square base of acrylic of 65 x 120 cm of external area, of 1 cm thick. Bars of Nylamid were also used for support. The used materials minimize the interaction with the electromagnetic fields that could be present.

3.1.3 Communication and control stage

The communication system is based on the use of two IBM compatible computers, one of them inside the anechoic chamber (remote) and the other one in the exterior (local). The communication between both computers was carried out through an optical fiber cable; the local PC operates the remote PC. The local PC controls all the equipment inside the chamber and captures all the acquired information. An optical fiber based net card was used to communicate the two computers (3C905B-FX(SC), 3Com 100Base-FX). It possesses a speed of transmission of 100 Mbps. The protocol used to communicate both computers was the TCP/IP. The package used for the communication was the Remote Administrator v2.0, which allows control all the devices and programs of the remote computer from the local computer. The analog signals coming from the eight conditioning circuits were digitized by means of a 12 bits analog to digital card (Lab PC-1200 A/I, National Instruments) that is installed in the remote computer. The communication system is based on the use of two IBM compatible computers, one of them inside the anechoic chamber (remote) and the other one in the exterior (local). The communication between both computers was carried out through an optical fiber cable; the local PC operates the remote PC. The local PC controls all the equipment inside the chamber and captures all the acquired information.

3.1.4 Visualization software

Software was developed in the LabWindows CVI programming language (National Instruments). By means of this software the user can control all the controllable devices that conforms the automated laboratory. Data like the radiation time and frequency, the distance between the radiator and the phantom, the quantity of temperature measurements in a plane of the phantom, the quantity of phantom mappings, sensor thermal response time, the elapsed time between each temperature mapping, among other, should be provided by the user. Once the experiment begins, the laboratory operates automatically and allows the user to visualize the results obtained in each temperature mapping. In Figure 4 the main screen of the program is shown.

3.2 Results

Calibration of the sensor was done using a thermostatic bath and a thermocouple thermometer with an accuracy of 0.1"C (TES-1310). Several calibration curves in different days were taken for the sensor, as shown in Figure 5. With these curves we evaluated the repeatability and the stability of the sensor. We also evaluated the thermal time constant of the sensor, resulting in a value of 1.9 s. As we see, the transmitted power increases with the temperature. While the temperature increases, the refraction index of the cladding decreases in a linear way. At low temperatures (below 5"C), the refraction index of the oil is higher than the refraction index of the core, and no power is transmitted through the fiber, because

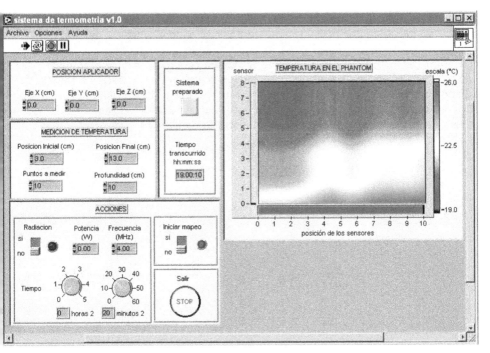

Fig. 4. Software main screen.

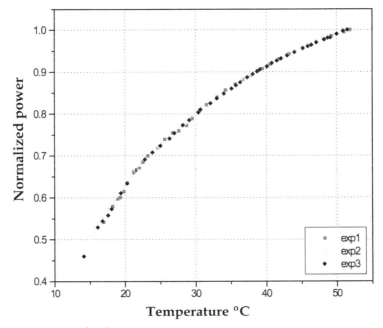

Fig. 5. Calibration curves for three experiments.

waveguide conditions are not satisfied. As the refraction index decreases the power transmitted increases, due to the increasing number of guided modes in the fiber. When the oil index equals the refraction index of the cladding, the transmitted power reaches a maximum value, and the curve shows a saturation effect. The sensor performance using microwaves was tested using a domestic microwave oven that operates at 2.45 GHz. The experimental setup was similar to the previously described. Instead the thermostatic bath, we have used the microwave oven, to heat water inside a plastic container. Figure 6 shows the achieved results, compared with a calibration curve obtained with the thermostatic bath.

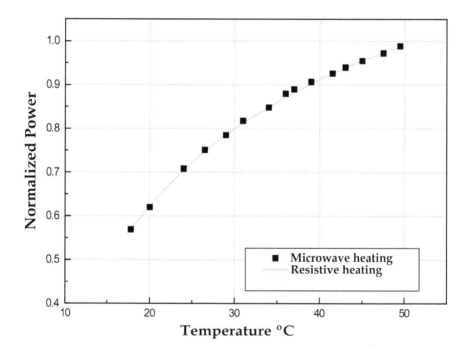

Fig. 6. Calibration curves for three experiments.

Figure 7 shows the fiber optic temperature sensor constructed for the hyperthermia laboratory.

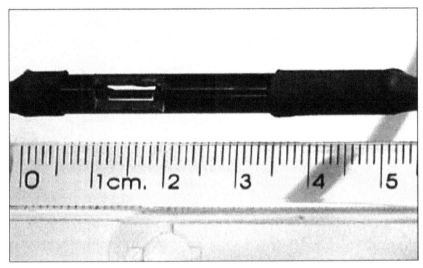

Fig. 7. Fiber optic temperature sensor constructed.

3.3 Discussion

A reasonable working range for the thermometry system is between 20°C to 35°C, as the phantom is heated about 3°C to 5°C from ambient temperature. According to the required precision, the thermometer should be accurate to within ±0.2°C at the: time of calibration. Additionally, a low thermal time constant and a low drift are required. The results obtained with the developed sensor show an appropriate performance for the selected application. The sensor was evaluated using microwaves to heat the bath in the operational interval of temperature. In this situation, we found that the sensor response still remains between the confidence band. The observed non-linearity is not an important issue, since we think to digitize the readings of multiple sensors with a personal computer. That implies the use of a calibration table to obtain the temperature values relative to the optical power detected. The reported results were obtained by means of optical instruments, but the idea is to use specifically designed circuits. For this reason, we are developing an optical source and optical detector circuits, to construct a thermometer with the presented sensor. As preliminary results, we have an optical source with an excellent stability (±0.0005 dB), thanks to the use of an optic feedback technique. That grants the required long term drift requirements of the thermometer. The optical components that we are using are the pair OPF1414-OPF2414 (Optek Inc.), which operate at a wavelength of 840 nm and possess ST ports for direct connection of the sensor. The procedure of construction of the sensor is simple and the required materials are relatively inexpensive. For example, the used fiber is a standard telecommunications grade patch-cord. The sensor could be used in other applications involving temperature measurements under strong electromagnetic fields, such as power transformers, microwave antennas, etc. The measurement range could be changed simply using oil with an appropriate refraction index. One of the drawbacks is the limited range within the sensor can make the measurements.

4. Fiber optics thermometry in microwave ablation

In this section, we present an experimental setup and results of temperature measurements during MWA using fiber optics thermometry in *ex vivo* tissue (Cepeda et al., 2011). Luxtron STB MAR'05 fiberoptic thermal probes were placed 5mm above the antenna slots longitudinally to measure real-time temperature during high hyperthermia experiment (Figure 8). The fiberoptic thermal probes are connected to a Luxtron 3300 Fluoroptic thermometer, which monitors the temperature and saves the data to a personal computer via a RS-232 serial cable.

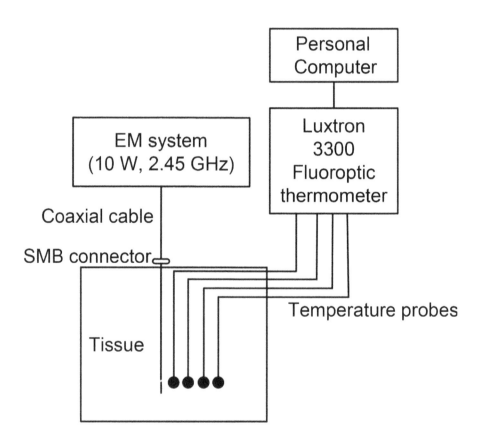

Fig. 8. Schematic of equipment and experimental setup. The antenna was completely inserted into homogeneous muscle tissue and connected to the EM system. Temperature sensors were 5 mm above the antenna slots longitudinally.

High temperature hyperthermia experiments were performed at 10 W for 3 min. The initial tissue temperature was 18 to 19 °C. Figure 9 shows the tissue with an ablated zone.

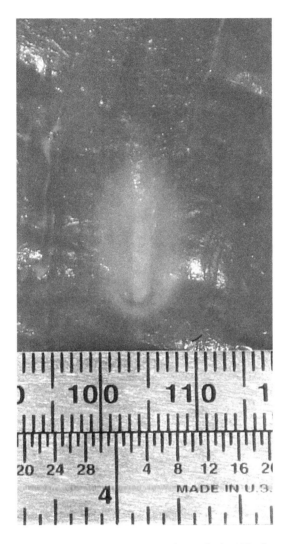

Fig. 9. A lesion created in an ex-vivo experiment at the end of a 180 s hyperthermia procedure.

The measured thermal histories at the specified locations during high hyperthermia therapy were shown in Figure 10. A fluoroptic thermometer was selected because its fiber optic temperature sensors have minimal disturbance on the antenna Specific Absorption Rate (SAR) and are unaffected by microwave radiation. Figure 11 shows a photo of *ex vivo* swine breast tissue temperature measurement, using one fiber optic during MWA.

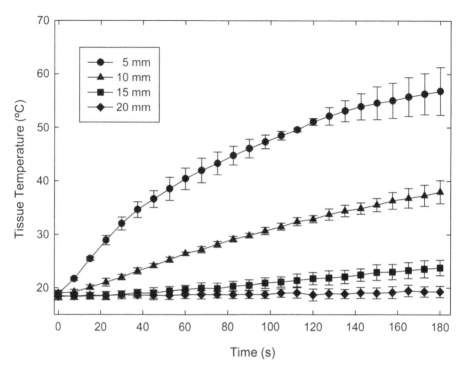

Fig. 10. A lesion created in an ex-vivo experiment at the end of a 180 s hyperthermia procedure.

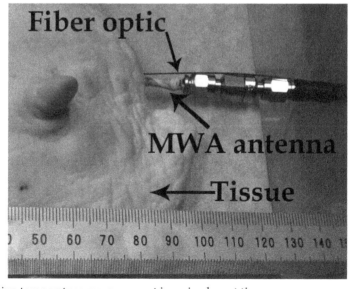

Fig. 11. Ex vivo temperature measurement in swine breast tissue.

5. Conclusion

All methods of hyperthermia involve transfer of heat into the body from an external energy source and are currently under study, including local, regional, and whole-body hyperthermia. Local hyperthermia is used to heat a small area. It involves creating very high temperatures that destroy the cells that are heated. Microwave ablation, like radiofrequency ablation, uses heating to cause tissue necrosis. RFA heats the tissue by electrical resistive heating but MWA works on a different principle. An antenna emits microwave radiation into the tissue; this results in excitation and oscillation of polar molecules and causes frictional heating. Frequently, temperature monitoring during hyperthermia cancer therapy has been achieved using interstitial thermocouples, thermistors, or fiber optic probes. The thermocouples and other conventional probes possess numerous advantages that make them attractive to be used in the characterization of temperature in hyperthermia. Among these advantages are a small size, accuracy, excellent reliability and low cost. Furthermore, this technology also possesses disadvantages, like metallic conductive components, shielding and plug wires. These metallic parts are often the reason of considerable errors in measurement of temperature when the probes are used for characterizing equipment of hyperthermia in phantoms exposed to electromagnetic radiations. These errors of measurement are caused by three phenomena, which can be presented alone or acting in combination: sensor heating due to the induced currents, disturbance of the electromagnetic field and electromagnetic interference. Thermometers based on optical fibers offer the advantage of not possessing metallic components, and therefore they do not to disturb the electromagnetic field. Fiber optic thermometers are used when electrical insulation and electromagnetic immunity are necessary. The most relevant application includes tissue-heating temperature control during radiofrequency or microwave hyperthermia cancer treatment.

6. References

Cepeda, M., Vera, A., Leija, L., Avila-Navarro, E. & Navarro, E. (2011). Coaxial Slot Antenna Design for Microwave Hyperthermia using Finite-Difference Time-Domain and Finite Element Method. *The Open Nanomedicine Journal*, Vol. 3, No. 1, (2011), pp. (2-9), ISSN 1875-9335

Chong, J., Leija, L., Posada, R., Fonseca, W. & Vera, A. (2000). Design and construction of an automated laboratory for electromagnetic and ultrasonic radiation in the study of hyperthermia in biological systems, *Proceedings of the 2000 Annual Fall Meeting of the Biomedical Engineering Society*, Seattle, WA, October 2000

Culshaw, B. (1999). Fiber-Optic Thermometers, In: *Measurement, Instrumentation, and Sensors Handbook*, John G. Webster, 32.11, CRC Press LLC, ISBN 978-0-8493-2145-0, USA

Goldberg, S. & Gazelle, D (2001). Radiofrequency tissue ablation: physical principles and techniques for increasing coagulation necrosis. *Hepatogastroenterology*, Vol. 48, No. 38, (2001), pp. (359-367), ISSN 0172-6390

Hildebrandt, B., Wust, P., Ahlers, O., Dieing, A., Srcenivasa, G., Kerner, T., Felix, R. & Riess H. (2002). The cellular and molecular basis of hyperthermia. *Critical Reviews in Oncology/Hematology*, Vol. 43, No. 1, (July 2002), pp. (33-56), ISSN 1040-8428

Izzo, F., Thomas, R., Delrio, P., Rinaldo, M., Vallone, P., DeChiara, A., Botti, G., D'Aiuto, G., Cortino, P. & Curley, S. (2001). Radiofrequency ablation in patients with primary breast carcinoma: a pilot study in 26 patients. *Cancer,* Vol. 92, No. 8, (October 2001), pp. (2036-2044), ISSN 1097-0142

Katzir, A., Bowman, H., Asfour, Y., Zur, A., & Valeri, C. (1989). Infrared fibers for radiometer thermometry in hypothermia and hyperthermia treatment. *IEEE Transactions on Biomedical Engineering,* Vol. 36, No. 6, (June 1989), pp. (634-637), ISSN 0018-9294

Mignani, A. & Baldini, F. (1996). Biomedical sensors using optical fibres. *Reports on Progress in Physics,* Vol. 59, No. 1, (January 1996), pp. (1-28), ISSN 0034-4585

Pennisi, C., Leija, L., Fonseca, W. & Vera, A. (2002). Fiber optic temperature sensor for use in experimental microwave hyperthermia, *Sensors, 2002. Proceedings of IEEE,* ISBN 0-7803-7454-1, Orlando, Florida, USA, June 2002

Saxena, I. & Hui, K. (2010). Polymer coated fiber Bragg grating thermometry for microwave hyperthermia. *Medical physics,* Vol. 37, No. 9, (September 2010), pp. (4615-4619), ISSN 0094-2405

Simon, C., Dupuy, D. & Mayo-Smith, W. (2005). Microwave Ablation: Principles and Applications. *RadioGraphics,* Vol. 25, No. suppl 1, (October 2005), pp. (S69-S83), ISSN 0271-5333

Vaguine, V., Christensen, D., Lindley, J. & Waltson, T. (1984). Multiple Sensor Optical Thermometry System for Application in Clinical Hyperthermia. *IEEE Transactions on Biomedical Engineering,* Vol. 31, No. 1, (January 1984), pp. (168-172), ISSN 0018-9294

van der Zee, J. (2002). Heating the patient: a promising approach?. *Annals of Oncology,* Vol. 13, No. 8, (August 2002), pp. (1173-1184), ISSN 1569-8041

van Esser, S., van den Bosch, M., van Diest, P., Mall, W., Borel Rinkes, I. & van Hillegersberg, R. (2007). Minimally Invasive Ablative Therapies for Invasive Breast Carcinomas: An Overview of Current Literature. *World Journal of Surgery,* Vol. 31, No. 12, (December 2007), pp. (2284-2292), ISSN 0364-2313

Wook, J., Jeong, K., Dong, H., Kyoung, W., Ji, Y., Jae, H., & Bongsoo L. (2009). Infrared Fiber-Optic Temperature Array Sensor for Radiofrequency Ablation, *Proceedings of Biomedical Engineering and Informatics, 2009. BMEI '09. 2nd International Conference on,* ISBN 978-1-4244-4132-7, Tianjin, October 2009

Wust, P., Hildebrandt, B., Sreenivasa, G., Rau, B., Gellermann, Riess H., Felix, R. & Schlag P. (2002). Hyperthermia in combined treatment of cancer. *The Lancet Oncology,* Vol. 3, No. 8, (August 2002), pp. (487-497), ISSN 1470-2045

10

White Light Sensing Systems for High Voltage Measuring Using Electro-Optical Modulators as Sensor and Recovery Interferometers

Josemir C. Santos[1], José C. J. Almeida[2] and Luiz P. C. Silva[1]
[1]University of São Paulo
[2]Optsensys
Brazil

1. Introduction

There are several known issues related to the use of electro-optical sensors as measuring devices. In applications intended for use in measuring currents and potentials in high voltage electric power systems, optical sensor systems must attend severe specifications, such as: large withstand voltage of the primary sensor, high accuracy (better than 0,5%, in case of revenue metering), long term stability of scale factor, large dynamic range (>50 dB, in some applications for protection purpose) and relatively wide frequency response range (of several kHz, in case that harmonics content evaluation is needed). However, the potential advantages arising from the use of optical fiber sensors as instrument transformers (ITs) in substitution to conventional electromagnetic transformers, may explain the intensive efforts that have been done during the last 30 years to develop optical ITs.

In a previous work (Santos, 2009), focused on the development of electro-optical sensors applied to direct measuring high voltages, in order to fulfill the requirements for ITs previously described, various sensing systems linked by optical fibers were presented. The progress in this field is commonly divided into three main approaches.

In the first approach, conventional transformers connected to primary electronic converters and communicated by means of optical fiber links to secondary converters have been used to improve the EMI toughness and to provide complete electrical isolation between the measured and the measuring systems. Typical examples of such approach (Crotti et al., 2006; Rahmatian, 2010) describe conventional ITs connected to primary electronic converters that generate modulated optical signals and provide their adequate coupling to optical fiber communication links.

In the second aproach, non-conventional transformers, also called as electronic transformers, have been developed using different types of sensors or transducers as measuring devices that are connected by optical fiber links to secondary converters. In representative examples of this approach (Bull et al., 2005; Mariscotti, 2009) Rogowski coils are used as sensor element.

The third approach consists in using optical sensors to directly measuring of voltage or current in high voltage systems. Passive optical sensors, such as electro-optical and magneto-optical modulators, are attractive for such purpose because all their components can be made with dielectric materials and have small dimensions. Since these sensors, generally, show low sensitivity to EMI and wide frequency response, they usually have much superior performance characteristics than the conventional transformers, and are capable to make better use of potentially excellent quality of the optical fiber links.

The linear electro-optic effect, also called Pockels effect, has been largely used to build electro-optical modulators that have many applications in communications, signal processing and sensing systems. These modulators can be used as sensors for measuring of electric field, current and voltage, among other quantities. The optical sensors used as ITs for measuring voltage and current are called Optical Voltage Transformers (OVTs) and Optical Current Transformers (OCTs), respectively.

The main problem in trying to apply electro-optical sensors to perform direct measuring of potentials at high voltage levels is their sensitivity to applied voltage, commonly too high in comparison with the voltage to be measured. Since in typical Pockels modulators the transmitted light intensity has a sinusoidal dependency with applied modulation voltage, as shown later in this chapter, there is a limit for holding a univoque relationship between the transmitted light intensity and the applied voltage, imposing a restriction in maximum value of applied voltage that can be actually measured.

A solution that has been adopted to allow measuring of higher voltages relies in using capacitive or resistive potential dividers for obtaining a sample of the high voltage to be measured and to apply this low voltage signal to the optical sensor. This solution limits the performance of the OVT to that of the potential divider, which is not always satisfactory. The most favorable solution, however, should be the direct application of high voltages to a properly designed Pockels modulator, avoiding the use of dividers and, therefore, making the best use of superior characteristics of optical sensors. To reach this goal, the Pockels modulator should be built in an appropriate configuration, in order to increase its dielectric withstand and reduce its sensitivity to suitable values. Several alternatives have been proposed to build optical fiber sensors appropriate for high voltage measuring systems, but a small number of them could lead to practical devices able to be used in field applications.

Another approach to solve the posed problem is the application of a different interrogation technique, instead of the polarimetric intensity modulation normally used in Pockels modulators, capable to extend the dynamic range of the sensor to a sufficiently high value.

Recently, the White Light Interferometry (WLI) technique has been used as an alternative to implement optical fiber sensor systems that can meet severe specifications, similar to the requirements of applications as in ITs. White light interferometric systems based on interferometers linked by optical fibers have been largely used for implementation of various high performance optical sensors on trial basis. The design of a specific sensor depends upon the particular requirements of its application as well as the physical quantity to be measured, which may be, for example, displacement, strain, temperature, etc. In such a kind of WLI sensor systems, a broadband light source is used to illuminate two interferometers that are connected in series by an optical fiber link. The application of this method can add a large dynamic range to the high accuracy, which is a typical advantageous characteristic of conventional interferometric sensor systems.

Following this approach, it is shown in this chapter how WLI technique can be applied to a Pockels voltage sensor, enhancing its performance while keeping its linearity and accuracy.

At first, the general concepts related to WLI are presented.

Next, it is introduced an application, originally proposed by the first author, in which the WLI is used together with two electro-optical modulators, based on the Pockels effect, for measuring high voltages directly. Since this approach has no moving parts, the frequency bandwidth achievable can be much higher than any mechanically or electronically scanned WLI system previously proposed.

Finally, practical realizations of ITs based on WLI sensor systems using electro-optical modulators as sensor and recovery interferometers are presented. Special attention is given to electronic signal processing techniques used to demodulate the optical signal available at the output of the sensor system. Two different techniques are depicted; both capable to perform measuring with compensation for fluctuations on optical average power delivered at the output of sensor system. Such fluctuations normally arise from variations in attenuations or variations in the output power of optical source, for example. Experimental investigations carried out with the sensor systems developed are presented, including experimental results obtained with both electronic signal processing techniques. At the end of chapter, a brief discussion on the future perspectives and under developing initiatives related to this field is addressed.

2. Principles of White Light Interferometry (WLI) sensing

In typical interferometric optical sensors, highly monochromatic (or long coherence length) light sources are commonly used, therefore the unambiguous measurement range is limited to one period of the interferometer transfer function and the basic unit of measurement is the source wavelength.

By the other hand, white light interferometry has, potentially, the capability of identifying the interference fringe order from the output fringe pattern by using a broadband optical source (Chen, 1992). As a result, the unambiguous range of the output signal is no longer limited to within half of a fringe, and an absolute phase measurement over a large operating range can be achieved. The output intensity pattern of a white light interferometer has a visibility profile determined by the low-coherence property of the broadband source used.

Since the output of an interferometer is, in theory, the Fourier transform of the source spectrum (Françon, 1967) and the Fourier transform a Gaussian function is also a Gaussian function, it follows that the normalized output intensity pattern of a white light interferometer is a cosine function modified by a Gaussian visibility profile, as shown in Fig. 1. This output intensity pattern is commonly also called as fringe pattern in optical engineering field.

The principle of WLI sensing may be explained by using Fig. 2, which shows a block diagram of a generic WLI sensor system.

In Fig. 2 it is shown that light emitted from a spectrally broadband source, for example a superluminescent diode (SLD), is coupled via a beam splitter into an optical fiber link connected to the sensor interferometer. The optical path difference (OPD) introduced by the

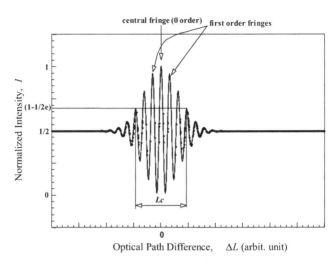

Fig. 1. Theoretical output intensity pattern of a white light interferometer.

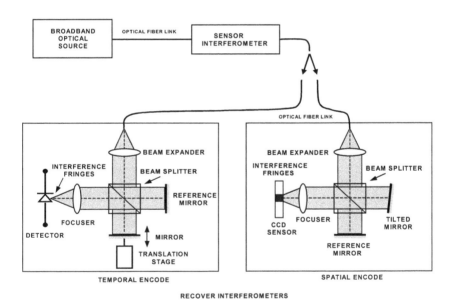

Fig. 2. Principle of operation of a white light interferometric sensor.

sensor interferometer, $\Delta L_s(X)$, is linearly sensitive to the external measurand X. If the OPD introduced by the sensing interferometer is greater than the coherence length of the source, L_c, no interference is observed at the output light of the sensor interferometer. When the output of sensor interferometer is linked to the input of a second interferometer, called as "recovery" interferometer, scanning of the path difference introduced by this interferometer, $\Delta L_r(Y)$, results in interference being observed at its output when $\Delta L_r(Y)$ approaches to $\Delta L_s(X)$.

Neglecting the spectral dependency of OPD, the behavior of light intensity at the output of sensor interferometer, I_s, as a function of light wavenumber, σ, can be described by the following relation (Lequine, 1990):

$$I_s(\sigma) = T_1 T_s I_o(\sigma)\{1 \pm K_s \cos[2\pi\sigma\Delta L_s(X)]\} \tag{1}$$

where T_1 is the transmission factor of the forward fiber link, T_s and K_s are the transmission factor and the visibility of sensor interferometer, respectively, and $I_0(\sigma)$ is the spectral intensity distribution of the light source. Analogously, the behavior of the recovery interferometer can be written as:

$$I_r(\sigma) = T_2 T_r I_s(\sigma)\{1 \pm K_r \cos[2\pi\sigma\Delta L_r(Y)]\} \tag{2}$$

where T_2 is the transmission factor of return fiber link and T_r and K_r are the transmission factor and the visibility of recovery interferometer, respectively. The OPD introduced by the recovery interferometer, $\Delta L_r(Y)$, is linearly sensitive to scanning quantity Y. The total intensity available for detection at the output of WLI system is calculated by integrating the previous equation over all the wavenumbers, and, if the source has a Gaussian spectral distribution, this integral becomes:

$$I = T_1 T_2 T_s T_r I_0 \left\{1 + K_{0s} K_{0r} e^{-\left[\pi\frac{(\Delta L_s - \Delta L_r)}{L_c}\right]^2} \cos\left[2\pi\frac{(\Delta L_s - \Delta L_r)}{\lambda_0}\right]\right\} \tag{3}$$

The resulting intensity signal, I, appears in eq. (3) as a cosine function modified by a visibility function, both dependent of the OPD difference, exactly as shown in Fig. 1. Both eq. (3) and Fig. 1 show that, at the matching point, the interferometer output is at the center of the brightest fringe. The order of this fringe is zero and it is called *central fringe*. The fringe-order identification is dependent on the identification of central fringe.

To achieve a precise measurement in a large dynamic range, the sensor system must, at first, correctly identify the central fringe and then measure accurately the OPD introduced by the recovery interferometer, $\Delta L_r(Y)$, necessary to lead the system to the matching point, i.e., the center of central fringe. There are several proposed methods to reach this aim, using different topologies and signal processing techniques. The two most commonly used OPD scanning techniques, the temporal encode and the spatial encode, are illustrated in Fig. 2.

Temporal encode techniques make use of mechanically tunable OPD interferometers to scan the fringe pattern and record it for further analysis. This method allows high accuracy and low signal-to-noise ratio (SNR) level, but it is normally very slow, since it uses moving parts.

In a system using electronically scanned (spatial encode) WLI, the interference fringes pattern is produced by superposing two expanded light beams. The resulting pattern is imaged over a CCD array. This method is much faster than the temporal encoding, but the fringe visibility is reduced due to spatial coherence mismatches and intrinsic noises of beam profile and CCD array (Ning et al., 1995).

In both cases, the signal processing techniques normally include digital treatment of the signal and the computation time may limit the response frequency bandwidth of the sensor system. Besides these two techniques, other electronically scanned methods using dual wavelength light sources and no moving parts can be applied to WLI sensors.

In this chapter a new approach for temporal encoded WLI sensor systems with no moving parts is proposed by using a Pockels electro-optic modulator as recovery interferometer. In such approach, the frequency bandwidth of the system can be significantly enhanced while keeping a large measuring range of operation.

3. WLI method applied to fiber optical linked electro-optical sensor systems

Recently, several WLI sensor systems have been proposed to measure different physical quantities (measurands). In Table 1 some examples are listed and information are given about the measurand, the encoding technique, the bandwidth and the dynamic range for each sensor.

Measurand	Encoding Technique	Bandwidth	Dynamic Range	Reference
Temperature	Temporal (fiber stretch)	30 Hz	20 ~ 70 °C	(Kim, 2008)
Position	Temporal (moving mirror)	< 1 Hz	0 ~ 7.2 mm	(Chen, 2010)
Displacement	Temporal (moving mirror)	~20 Hz	0 ~ 1050 nm	(Liao, 2010)
Micro-Strain	Temporal (moving lens)	1 Hz	0 ~ 350 με	(Velosa, 2011)
Temperature	Temporal (moving lens)	> 60 Hz	35 ~ 75 °C	(Yu, 2005)
Displacement	Spatial (electronically scanned)	~500 Hz	0 ~ 20 μm	(Murtaza, 1991)

Table 1. Examples of WLI sensor systems

From table 1, it is possible to observe that typical WLI sensor systems presented show dynamic ranges that can be considered larges, when compared with the ones achievable by normal interferometric systems in same conditions. However, in all cases, relatively narrow bandwidths are observed, preventing the use of mentioned systems in applications intended for measurement of fast varying measurands.

In alternate current (ac) electrical power systems the main electrical measurands of interest are voltage and current, and the lowest nominal frequency of these quantities is 50 Hz. Therefore, only the example shown in last line of table 1 could be applied to measure such quantities with significant information about their harmonic content. Furthermore, larger bandwidth may be required from a sensor system to be applicable as IT for protection purposes, since they must to respond to transitory phenomena, such as short circuits or lightning and switching impulses, which superpose overvoltage and overcurrent variations much faster than the period associated to the nominal frequency. Consequently, practical application of WLI technique in optical ITs demands advances in the fields of encoding techniques and signal processing schemes, to allow production of sensor systems able to provide bandwidths sufficiently larges to attend requirements of defined accuracy classes, which will depend on the foreseen use of IT for metering or protection.

White Light Sensing Systems for High Voltage Measuring Using Electro-Optical Modulators as Sensor and
Recovery Interferometers

203

Recently a new encoding technique based on electro-optical interferometers has been developed to overcome both mentioned limitations, in dynamic range and bandwidth, for the specific case of OVT to be used for measuring in high voltage electric power systems.

Most part of electro-optical sensors utilized in OVTs makes use of electro-optic modulators based on Pockels effect configured as birefringence interferometers. In this case, there is a limitation in dynamic range of measured voltage, which is a typical disadvantage arising from the polarimetric method used for measuring the electrically induced phase retardation due to the Pockels effect, since normal interferometric methods are useful only to measure OPD in the range of one wavelength. However, as described earlier, WLI sensor systems are based on interferometers linked in series by optical fibers. Therefore, by linking in series two properly designed Pockels modulators configured as birefringence interferometers, one as voltage sensor and other as recovery interferometer, it is possible to set up a WLI sensor system.

The working principle of a Pockels modulator, represented as shown in Fig. 3, can be described as a two-beam birefringence interferometer by assuming that the components of incident light, which are polarized parallel to x and y axes, are two independent beams having intensities I_x and I_y, respectively. When light is crossing the Pockels crystal, both components travel along a path of same length, L, which is the length of crystal, but each one is affected by a different refractive index. The difference between these two refractive indices, $\Delta n = n_y - n_x$, is the birefringence of medium. The light emerging from the crystal crosses a second polarizer, called as analyzer, which is oriented at 45° with respect to

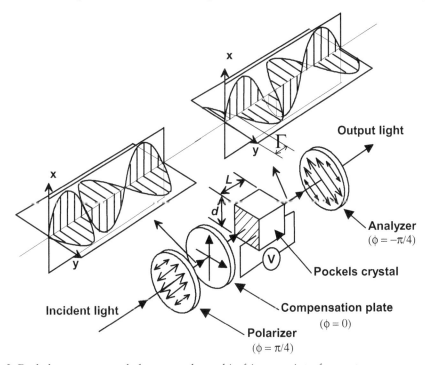

Fig. 3. Pockels sensor regarded as a two-beam birefringence interferometer.

polarization directions of both components. Since the analyzer transmits only the part of light intensity of each component that is parallel to its polarization direction, interference of the two light components occurs at its output.

The key part in designing a Pockels modulator is the calculus of its birefringence. The general procedure involved in this task depends on several parameters, such as: the specific crystal selected (including its optical and electrical characteristics, geometry and dimensions), the configuration chosen (longitudinal or transverse) and the relative orientation among the propagation direction, the electric filed direction and the crystal's optical axes directions.

For example, using an electro-optical crystal with cubic lattice (of $\overline{4}3m$ symmetry group, for instance), in a polarimetric modulator in longitudinal configuration as shown in Fig. 3, where polarizer and analyzer polarization directions are crossed to each other, the propagation direction is parallel to an optical axis of crystal, and considering that applied voltage V generates an uniformly distributed electric field, E, inside crystal, which implies that $E = V/L$, the electricaly induced birefringence, Δn_e, becomes (Yariv & Yeh, 1984):

$$\Delta n_e(V) = n_o^3 r_{41} E = n_o^3 r_{41} \frac{V}{L} \tag{4}$$

where n_O is the ordinary refractive index and r_{41} is, in this case, the only relevant electro-optic coefficient of crystal.

The electrically induced phase retardation, Γ_e, between the two orthogonal light components is:

$$\Gamma_e = \frac{2\pi}{\lambda} \Delta n_e \times L \tag{5}$$

where λ is the wavelength of light inside crystal when there is no applied voltage.

In this case, the half-wave voltage, which is defined as the value of V that makes $\Gamma_e = \pi$, is given by (Almeida, 2001):

$$V_\pi = \frac{\lambda}{2n_o^3 r_{41}} \tag{6}$$

The total phase retardation, Γ_t, is the summation of the electrically induced phase retardation, Γ_e, with a fixed contribution of compensation plate, $\Delta\phi_c$, and can be written as:

$$\Gamma_t = \Delta\phi_c + \frac{2\pi}{\lambda} \Delta n_e \times L = \Delta\phi_c + \frac{2\pi}{\lambda} n_o^3 r_{41} V \tag{7}$$

The total optical phase retardation can be also given in terms of V_π by:

$$\Gamma_t = \Delta\phi_c + \pi \frac{V}{V_\pi} \tag{8}$$

By using eq. (8) the optical path difference, ΔL, correspondent to this optical phase retardation is given by:

$$\Delta L = \frac{\lambda}{2\pi}\Gamma_t \tag{9}$$

Applying eq. (7) to this relation, the total path difference becomes

$$\Delta L = \Delta L_c + n_o^3 r_{41} V \tag{10}$$

where $\Delta L_c = \frac{\lambda}{2\pi}\Delta\phi_c$ is the OPD introduced by the compensation plate.

From eq. (10), it is found that the total OPD in this interferometer is independent of wavelength and is linearly dependent on applied voltage. Since the two beams have crossed polarization directions they cannot interfere with other. The recombination of beams is made by the analyzer, which actually is a polarizer with its polarization direction oriented at -45° with respect to x axis. Since these two beams have the same intensity ($I_x = I_y$) and only half of intensity of each beam is transmitted by the analyzer, the visibility of its central fringe is $K_{0s} = 1$, and the transmission factor of this sensing interferometer is:

$$T_s = \frac{1}{2}\alpha_s \tag{11}$$

where α_s is the attenuation introduced in the transmitted light by the sensing interferometer due to factors such as reflections, scattering, misalignments, absorption, etc.

The behavior of this interferometer is given, from eq. (2), by:

$$I_s(\sigma) = T_1 \alpha_s \frac{1}{2} I_0(\sigma)\left\{1 \pm \cos\left[2\pi\sigma\left(\Delta L_c + n_o^3 r_{41} V\right)\right]\right\} \tag{12}$$

If a second Pockels birefringence interferometer is linked in series with the sensing interferometer described above, the overall behavior of the white light interferometric sensor system obtained is given, analogously to eq. (3), by:

$$I = \frac{1}{4}\alpha_s\alpha_r T_1 T_2 I_0\left\{1 + \frac{1}{2}e^{-\left\{\left[(\Delta L_{cs} - \Delta L_{cr}) + n_o^3 r_{41}(V_s - V_r)\right]\frac{\pi}{L_c}\right\}^2}\cos\left[2\pi\frac{(\Delta L_{cs} - \Delta L_{cr}) + n_o^3 r_{41}(V_s - V_r)}{\lambda_0}\right]\right\} \tag{13}$$

where indices s and r correspond to values related to sensor and recovery interferometer, respectively, and L_c is the coherence length of source, given by (Yu, 2005):

$$L_c \cong \frac{\lambda_0^2}{d\lambda} \tag{14}$$

where λ_0 is the central wavelength of low-coherence source in vacuum and $d\lambda$ is the spectral width of source.

In order to operate the sensor system near to central fringe, both OPDs introduced by retardation plates must to be equal. In this situation, $\Delta L_{cs} - \Delta L_{cr} = 0$ and eq. (13) becomes:

$$ I = \frac{1}{4}\alpha_s\alpha_r\,T_1 T_2\,I_0 \left\{ 1 + \frac{1}{2}e^{-\left\{\left[n_o^3 r_{41}(V_s - V_r)\right]\frac{\pi}{L_c}\right\}^2}\cos\left[2\pi\frac{n_o^3 r_{41}\left(V_s - V_r\right)}{\lambda_0}\right]\right\} \tag{15} $$

It is found from last equation that, for a system built as described, the voltage applied to recovery interferometer, V_r, must to be equal to the voltage to be measured, V_s, in order to reach the maximum light intensity at system output. Therefore, by observing the output light intensity, I, meanwhile V_r is changed and measuring V_r when I is at maximum, it is possible to know directly the value of V_s at that moment.

The measuring method just described corresponds to a hybrid encoding technique, since it can be considered as temporal, once it could be necessary to scan periodically V_r to get one measured value of V_s per period and, also, it can be seem as electronically scanned, once there are no moving parts in the system and the OPD scanning is performed by means of an electro-optical device.

The exact classification of this technique, in fact, depends on how the electronic signal processing of the sensor system is implemented. The behavior and performance characteristics of the sensor system are ultimately defined by the association of encoding technique with the signal processing technique.

4. Application of WLI technique to a high voltage Pockels sensor system

Considering the application of a WLI sensor system as high voltage OVT, the method just described is not useful because it imposes the need to apply to recovery interferometer the same high voltage that is applied to the sensor interferometer, which is meaningless.

There are, at least, two possibilities for introducing a measurement scale factor in the sensor system, in order to reduce the maximum value of scanning voltage necessary to find the central fringe position. The first alternative consists of using different crystals in sensor and recovery interferometers. Choosing a crystal to be used in recovery interferometer with a product $n_{or}^3 r_{41r}$ greater than the product $n_{os}^3 r_{41s}$ of crystal used in sensor interferometer, the voltage V_r necessary to matches the OPD introduced by sensor interferometer will be smaller. In this case, eq. (15) becomes:

$$
\begin{aligned}
I &= \frac{1}{4}\alpha_s\alpha_r\,T_1 T_2\,I_0 \left\{ 1 + \frac{1}{2}e^{-\left[\left(n_{os}^3 r_{41s}V_s - n_{or}^3 r_{41r}V_r\right)\frac{\pi}{L_c}\right]^2}\cos\left[2\pi\frac{n_{os}^3 r_{41s}V_s - n_{or}^3 r_{41r}V_r}{\lambda_0}\right]\right\} = \\
&= \frac{1}{4}\alpha_s\alpha_r\,T_1 T_2\,I_0 \left\{ 1 + \frac{1}{2}e^{-\left\{\left[n_{os}^3 r_{41s}(V_s - fV_r)\right]\frac{\pi}{L_c}\right\}^2}\cos\left[2\pi\frac{n_{os}^3 r_{41r}\left(V_s - fV_r\right)}{\lambda_0}\right]\right\}
\end{aligned}
\tag{16}
$$

where $f = \dfrac{n_{or}^3 r_{41r}}{n_{os}^3 r_{41s}}$ is the scale factor.

The second alternative to introduce a scale factor is making use of a configuration in the Pockels modulator applied as sensor interferometer different than the configuration used in the modulator used as recovery interferometer. For example, a transverse modulation configuration can be applied in recovery interferometer while longitudinal modulation configuration is applied in sensor interferometer. In such a case, eq. (15) becomes:

$$
I = \frac{1}{4}\alpha_s\alpha_r T_1 T_2 I_0 \left\{ 1+\frac{1}{2}e^{-\left\{\left[n_o^3 r_{41}\left(V_s-\frac{L}{d}V_r\right)\right]\frac{\pi}{L_c}\right\}^2} \cos\left[2\pi\frac{n_o^3 r_{41}\left(V_s-\frac{L}{d}V_r\right)}{\lambda_0}\right] \right\}
\tag{17}
$$

where d is the thickness of Pockels crystal used in the recovery interferometer and the scale factor is L/d. In order to describe in a general form the scale factor arising from the use of different electro-optic modulators in WLI sensor systems, eq. (15) may be represented in terms of V_π as follows:

$$
I = \frac{1}{4}\alpha_s\alpha_r T_1 T_2 I_0 \left\{ 1+\frac{1}{2}e^{-\left[\left(\frac{\lambda_0}{2V_{\pi s}}V_s-\frac{\lambda_0}{2V_{\pi r}}V_r\right)\frac{\pi}{L_c}\right]^2} \cos\left[\pi\left(\frac{1}{V_{\pi s}}V_s-\frac{1}{V_{\pi r}}V_r\right)\right] \right\} =
$$

$$
= \frac{1}{4}\alpha_s\alpha_r T_1 T_2 I_0 \left\{ 1+\frac{1}{2}e^{-\left\{\left[\frac{\lambda_0}{2L_c}\frac{\pi}{V_{\pi s}}(V_s-f_\pi V_r)\right]\right\}^2} \cos\left[\frac{\pi}{V_{\pi s}}(V_s-f_\pi V_r)\right] \right\}
\tag{18}
$$

where $V_{\pi r}$ and $V_{\pi s}$ are half wave voltages of Pockles modulators used as sensor and recovery interferometers, respectively, and the scale factor, expressed in terms of V_π, is: $f_\pi = \frac{V_{\pi s}}{V_{\pi r}}$.

The advantage in using f_π is that, regardless of all characteristics of any specific electro-optical modulator selected to be used as sensor or recovery interferometer, it is enough to calculate $V_{\pi r}$ and $V_{\pi s}$ to determine the optical behavior of WLI sensor system. Therefore, the procedure to design a high voltage OVT using a WLI sensor system as described must include the selection of a scale factor in such a way that the scanning voltage becomes suitable to be driven and measured by an electronic signal processing system.

4.1 Pockels modulators as high voltage sensor interferometers

Recently, different possibilities for development of Pockels modulators specially designed for direct measurement of high voltages have been investigated. Among these possibilities, a particular approach comprises an unusual configuration of Pockels modulator in which the electro-optic crystal is divided in several slices (Santos et al., 2000). This kind of modulator, which is called multi-segmented Pockels sensor, allows V_π to be defined to arbitrarily high value. Several prototype versions of devices based on such approach were already built and tested, demonstrating their suitability to be applied as primary sensor in high voltage OVTs.

A basic version of high voltage Pockels sensor in longitudinal modulation configuration consists in an arrangement of one piece of electro-optic Pockels crystal inserted between two aluminum electrodes, which are separated and supported by acrylic tubes. In a prototype formerly built, the electro-optic crystal used was a piece of $Bi_3Ge_4O_{12}$ (BGO) in parallelepiped shape (1x5x100 mm) encapsulated in an acrylic cylinder. Since BGO exhibits $n_o = 2.098$ and $r_{41} = 1.03\times10^{-12}$ (m/V), and using a SLD operating in $\lambda = 1.321$ µm as light source in the high voltage sensor described, the half-wave voltage calculated using eq. (6) is: $V_\pi \cong 69.4$ kV.

In order to increase V_π value, a high voltage Pockels sensor using longitudinal modulation in a multi-segmented configuration was developed to be used as sensor interferometer in an OVT prototype (Santos, 1996). Such sensor is composed by electro-optic crystals and pieces of dielectric material acting as mechanical supports that maintain the crystals spaced by gaps, as shown in Fig.4.

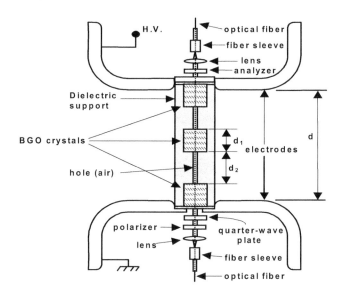

Fig. 4. Multi-segmented high voltage electro-optical sensor.

A simple method to make an estimation of V_π voltage for multi-segmented sensors is achieved by using the expressions (Santos, 1996):

$$V_\pi = \frac{(\lambda.K)}{2.n_0^3.r_{41}\ n.d_1} \tag{19}$$

$$K = \left[n.d_1 + \frac{\varepsilon_1}{\varepsilon_2}(n-1).d_2 \right] \tag{20}$$

where n is the number of crystal pieces, d_1 is their thickness, d_2 is the thickness of gaps, ε_1 is the permittivity of crystal and ε_2 is the permittivity of gas medium filling the gaps.

Although eq. (20) is a good approximation when crystal pieces are thin ($\cong 1$ mm), it does not remain valid when the thickness d_1 increases, due to non-uniformities of electric field inside the pieces. An equivalent value of K that allows a better estimation of V_π can be found from the voltage applied to electrodes and the average value of electric field in crystal pieces by:

$$K = \frac{V}{\overline{E}} \qquad (21)$$

$$\overline{E} = \frac{1}{n}\sum_{i=1}^{n}\overline{E}_i \qquad (22)$$

The average values of electric fields inside the crystals, $\overline{E}_i$, were computed by applying a 2D multiphysics simulation software based on finite elements method (FEM), which allows the exploration of axial symmetry presented by the sensor (Rubini et al., 2004). This kind of software has been used in analysis by FEM simulation of several electro-optic components, including integrated optics devices (Franco et al., 1999).

For example, using 3 pieces of BGO having d_1 = 6,26 mm, placed in a sensor as shown in Fig. 4, the calculated value using FEM simulation is: $V_\pi \cong 421,9$ kV. The experimentally measured value, using a prototype built with same configuration and materials above described, was: $V_\pi \cong 439,6$ kV. The good agreement between calculated and measured V_π values indicates that multi-segmented approach is suitable for obtaining Pockels modulators appropriate for direct measurement of high voltages.

4.2 Recovery interferometer

In order to allow the identification of the central fringe, a Pockels modulator suitable to be used as recovery interferometer in a WLI sensor system must to be able to reconstruct a significant part of the fringe pattern. Since each fringe is within an interval of 2π radians of phase difference and V_π is the voltage necessary to produce π radians of phase difference in a Pockels modulator, it follows that a Pockels recovery interferometer must have V_π as small as possible. Even though in a particular WLI sensor system multi fringe recovering is not required, a small V_π value is favorable for the electronic signal processing system, which is responsible for driving the scanning voltage V_r.

Reduction of V_π in electro-optic modulators can be achieved, for example, by adopting a transverse modulation configuration. In this case, the relation between electric field and applied voltage is given by $E = V/d$, and, from eq. (4) and eq. (5), V_π becomes:

$$V_\pi = \frac{\lambda}{2n_o^3 r_{41}}\frac{d}{L} \qquad (23)$$

To achieve smaller values of V_π, the ratio d/L must be minimized. Using a piece of BGO, in parallelepiped shape with dimensions of d = 1 mm and L = 100 mm, as electro-optic crystal in a prototype of recovery interferometer, and the same light source described before, the half-wave voltage calculated using Eq. (23) is $V_\pi \cong 694$ V. Although such value can be considered satisfactory for some applications, it is still relatively high.

An alternative to reduce V_π even more, as mentioned before, is to make use of materials with a product $n_{or}^3 r_{41r}$ greater than the product $n_{os}^3 r_{41s}$ of crystal used in sensor interferometer. It is possible to find crystals like Barium Strontium Niobate (BSN) that presents Pockels coefficients many times greater than that of BGO, which may allow obtaining much smaller V_π, as is desirable. Another possibility to obtain electro-optical modulators with low V_π is using integrated optical devices, which typically show half-wave voltages around 5 V. However, since there are no integrated optical devices suitable for WLI applications commercially available, development efforts must be done to make feasible this possibility.

5. Signal processing techniques applied to WLI high voltage sensor system

As stated earlier, behavior and performance characteristics of a WLI sensor system, as dynamic range, accuracy, resolution, bandwidth, thermal stability, etc., are all dependent on the association of an encoding technique with a signal processing technique.

The presented encoding technique based on Pockels modulators linked by optical fibers may, potentially, allow a WLI sensor system applied as OVT to have large bandwidth, high accuracy and fine resolution, among other desirable characteristics. However, since Pockels modulators can be seen as polarimetric interferometers, high voltages measured are traduced in terms of optical intensity at their outputs. Therefore, the measurement becomes dependent on link losses, which can change unpredictably in time.

Recently, applications of specially designed electronic signal processing systems have been proposed as options to overcome such problem (Almeida, 2001; Santos, 1996). The main idea behind these signal processing techniques is to take advantage of a particular property arising in WLI sensor systems when fixed OPDs introduced by compensation plates are larger than coherence length of the light source. In this case, from eq. (12) it is possible to observe that, since visibility approaches to zero, the information in the output of sensor interferometer is encoded only in the spectrum of light, suggesting that a measurement can be done independently of optical intensity present in the optical fiber link output.

In WLI systems using sensor and recovery interferometers connected in series, both introducing identical fixed OPDs, the information is translated back to the amplitude domain, as shown by eq. (13), and measuring can be done in a large bandwidth, but becomes again dependent on the optical intensity at output of system. To achieve a WLI sensor system that provides measurements independent on optical intensity it is necessary to develop electronic signal processing techniques specially designed to compensate optical intensity fluctuations not related to the measurand.

Following, two topologies of WLI sensor systems, using different encoding techniques associated with their specific signal processing techniques, both able to offer a.c. high voltage measurements compensated for optical intensity fluctuations, are presented. Prototypes were built and tested, and experimental results are presented and commented.

5.1 Peak and valleys values detection technique

5.1.1 Principle of operation

The strategy adopted in this signal processor approach is based on the precise detection of peaks and valleys values present in the output signal of a modulated recovery interferometer, which contains desired information about the ac high-voltage applied to the specially designed Pockels modulator used as sensor interferometer.

The analysis of this patent pending signal processing technique may start by the photo-current, $I_d(t)$, detected at output of optical fiber emerging from recovery interferometer, which is proportional to optical intensity, $I(t)$, as shown in Fig.5 (Almeida & Filho, 2003).

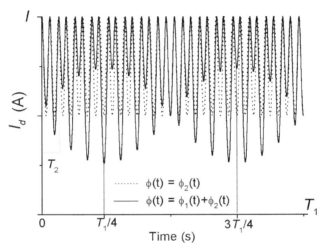

Fig. 5. Photo-current, I_d (t), for a period T_1 of the phase of modulation voltage applied to the sensor interferometer.

The behavior of $I_d(t)$ is shown in Fig. 5 for a particular condition, characterized by the phases of modulating voltages $\varphi_1(t) = \varphi_1 \sin \omega_1 t$ and $\varphi_2(t) = \varphi_2 \sin \omega_2 t$, applied, respectively, to sensor and recovery interferometers. To better explain the principle of operation, it were adopted $\omega_2 = 16\omega_1$, $\varphi_1(t) = 0.5$ radian, $\varphi_2(t) = \pi/2$ radians. $T_1 = 1/f_1$ is the period of fundamental frequency f_1 of $\varphi_1(t)$ and $\omega_2 = 1/T_2$ ($T_2 = T_1/16$).

In Fig. 5 it is illustrated that:

- the valley values change with the presence of phase modulation $\varphi_1(t)$ due to the ac high voltage applied to the sensor interferometer,
- these changes occur at the instants $T_2/4$ and $3T_2/4$ of each cycle of phase modulation $\varphi_2(t)$ applied to the recovery interferometer. This deviation of valley values occurs alternately bellow and above the reference value obtained without the phase modulation, $\varphi_1(t)$,
- the peak values don't change with the presence of phase modulation $\varphi_1(t)$ due to the ac high-voltage applied to the sensor interferometer. However, their positions change and occur symmetrically to the instant $T_2/2$ at each cycle of phase modulation, $\varphi_2(t)$.

In order to clarify the topics above, in Fig.6 it is shown a zoom in time, detailing one cycle of the voltage $V_d(t)$, present at the output of a transimpedance amplifier, used to convert the photocurrent $I_d(t)$ to voltage. An ideal transimpedance amplifier built in a classical configuration presents an output voltage $V(t)$ given by:

$$V(t) = R_f I(t) \tag{24}$$

were R_f is the feedback resistance, which is numerically equal to the transimpedance gain, G.

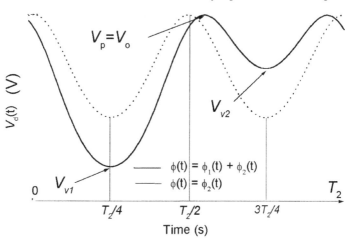

Fig. 6. Magnification in time showing output voltage at the output of transimpedance amplifier in a period of T_2.

In Fig.6 it is shown the occurrence of:

- a valley value V_{v1} and a valley value V_{v2}, bellow and above, respectively, reference value, represented by the dotted line, obtained for $\varphi_1(t) = 0$ radian,
- a peak value occurring after $T_2/2$.

The photo current, can be expressed by (Santos et al., 2002):

$$I_d(t) = \frac{I_o}{2}\left\{1 + \cos\left[\varphi_1(t) + \varphi_2 \sin \omega_2 t\right]\right\} \tag{25}$$

where I_o is its maximum value. $I_d(t)$ can also be expressed using Bessel functions as:

$$I_d(t) = I_o\left\{1 + \frac{1}{2}\left[J_0(\varphi_2) + 2\sum_n J_{2n}(\varphi_2)\cos(2n\omega_2 t)\right]\cos\left[\varphi_1(t)\right]\right\} +$$
$$- I_o\left\{\sum_n J_{2n-1}(\varphi_2)\sin\left[(2n-1)\omega_2 t\right]\sin\left[\varphi_1(t)\right]\right\} \tag{26}$$

where $n \in N^*$, and J_n denotes the nth order Bessel function of first kind (Almeida, 2001).

From eq. (25) and eq. (26) it is obtained the expression for the low frequency components of $I_d(t)$, labeled as $I_{lf}(t)$ and characterized by $\omega \langle\langle \omega_2 - \omega_1$, as:

$$I_{lf}(t) = I_o\left\{1 + \frac{1}{2}J_0(\varphi_2)\cos\left[\varphi_1(t)\right]\right\} \tag{27}$$

Likewise, from eq. (24) and eq. (27) the low frequency components of $V_d(t)$, termed as $V_{lf}(t)$, can be expressed as:

$$V_{lf}(t) = V_o\left\{1 + \frac{1}{2}J_0(\varphi_2)\cos\left[\varphi_1(t)\right]\right\} \qquad (28)$$

Based on eq. (24) and eq. (26) the high frequency components of $V_d(t)$ labeled as $V_{hf}(t)$ and identified by $\omega_2 - \omega_1 \rangle\rangle \omega_1$ can be written as:

$$V_{hf}(t) = V_o\left(\frac{1}{2}\left\{\cos\left[\varphi_1(t) + \varphi_2\sin\omega t\right] - J_0(\varphi_2)\cos\left[\varphi_1(t)\right]\right\}\right) \qquad (29)$$

From eq. (29) results that, the valley values of $V_{hf}(t)$, from now on termed $V_{v1}(t)$ and $V_{v2}(t)$, can be expressed in each cycle of $\varphi_2(t)$ as:

$$V_{v1} = \frac{V_o}{2}\left\{\cos\left[\varphi_2 + \varphi_1(t)\right] - J_0(\varphi_2)\cos\left[\varphi_1(t)\right]\right\} \qquad (30)$$

$$V_{v2} = \frac{V_o}{2}\left\{\cos\left[\varphi_2 - \varphi_1(t)\right] - J_0(\varphi_2)\cos\left[\varphi_1(t)\right]\right\} \qquad (31)$$

A straightforward substitution of trigonometric identities in eq. (30) and eq. (31), gives:

$$V_{v1}(t) = \frac{V_o}{2}\left\{\cos\varphi_2\cos\left[\varphi_1(t)\right] - \sin\varphi_2\sin\left[\varphi_1(t)\right]\right\} - \frac{V_o}{2}\left\{J_0(\varphi_2)\cos\left[\varphi_1(t)\right]\right\} \qquad (32)$$

$$V_{v2}(t) = \frac{V_o}{2}\left\{\cos\varphi_2\cos\left[\varphi_1(t)\right] + \sin\varphi_2\sin\left[\varphi_1(t)\right]\right\} - \frac{V_o}{2}\left\{J_0(\varphi_2)\cos\left[\varphi_1(t)\right]\right\} \qquad (33)$$

In order to maximize the influence of $\varphi_1(t)$ in the difference between $V_{v1}(t)$ and $V_{v2}(t)$, one should set $\varphi_2 = \pi/2$ radians as peak value of phase modulation $\varphi_2(t)$. Such a substitution in eq. (32) and eq. (33) results in:

$$V_{v1}(t) = \frac{V_o}{2}\left\{-\sin\left[\varphi_1(t)\right] - J_0(\pi/2)\cos\left[\varphi_1(t)\right]\right\} \qquad (34)$$

$$V_{v2}(t) = \frac{V_o}{2}\left\{\sin\left[\varphi_1(t)\right] - J_0(\pi/2)\cos\left[\varphi_1(t)\right]\right\} \qquad (35)$$

From eq. (34) and eq. (35), $\varphi_1(t)$ can be written as:

$$\varphi_1(t) = \arcsin\left[\frac{V_{v2}(t) - V_{v1}(t)}{(V_o/1,5)}\right] \qquad (36)$$

Strictly speaking, the parameter V_o changes with fluctuations in the intensity of the light incident on the photodiode and should be written as $V_o(t)$. In spite of that, eq. (36) shows that, from a theoretical point of view, these fluctuations have no influence on the measured value of $\varphi_1(t)$. From (36), in the interval $0 \le \varphi_2 \le \pi$, $\varphi_2 = \pi/2$ radians is the peak value of the phase modulation, $\varphi_2(t)$, which:

- maximizes the differential sensitivity between the valley values $V_{v1}(t)$ and $V_{v2}(t)$ as regard to $\varphi_1(t)$;
- minimizes the dependence of scale factor with respect to time fluctuations in φ_2, since that $\partial \varphi_2(t)/\partial t = 0$ for this point of operation;
- maximizes the dynamic range for the measurement of $\varphi_1(t)$, i.e., $|\varphi_1(t)|_{max} = \pi/2$ radians.

The presented technique allows keeping control of phase modulation depth, φ_2, by means of:

$$\varphi_2 = \arcsin\left[\frac{V_{v2}(t) + V_{v1}(t)}{(V_o/1,5)}\right] \tag{37}$$

likewise as provided by other techniques proposed for processing interferometric signals, based on division of RMS values of some of its harmonics or zero-crossings (Burns, 1994; Tselikov et al., 1998).

A blocks diagram of electronic signal processor developed for demodulating ac high voltages applied to OVT prototype is shown in Fig.7.

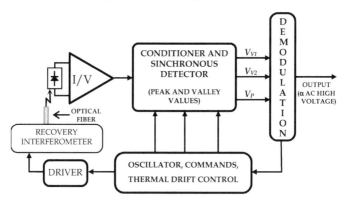

Fig. 7. Blocks diagram of the signal processor applied to the OPT.

The synchronous detection of consecutive peak and valleys values the transimpedance amplifier output yields, according to eq. (37), the demodulation of ac high voltage applied to WLI high voltage sensor system.

5.1.2 Experimental results

Aiming the development of a highly precise OVT based on WLI technique, a prototype of optical fiber sensor system using a Pockels modulator as recovery interferometer was built (Rubini et al., 2004), as shown in Fig. 8.

The sensor system was integrated to the electronic signal processor and the whole set was subjected to ac high-voltage tests conducted to determine the overall performance with

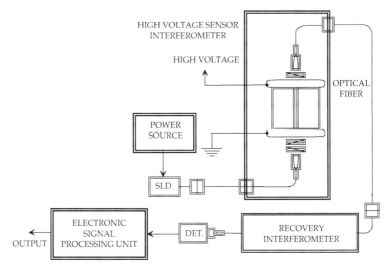

Fig. 8. Blocks diagram of OVT prototype using WLI configuration.

respect to the voltages applied. In Fig.9(a) it is illustrated the OVT used for the essays up to 10 kV_{RMS} applied to the sensor at the Optical Sensors Laboratory (LSO) of University of São Paulo (USP). The sensor interferometer is connected to a conventional instrument voltage transformer, rated 80.5 kV primary, 115/230V secondary, 7.5 kVA. Fig. 9(b) illustrates the OVT prototype at

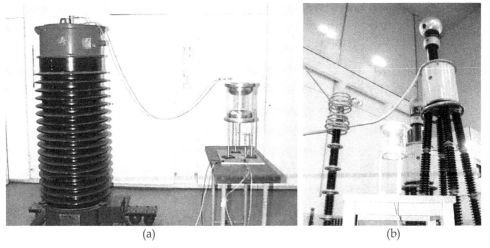

(a)   (b)

Fig. 9. Sensor interferometer of the OVT prototype: (a) connected to the voltage transformer at Optical Sensors Laboratory (LSO-USP) and (b) connected to the cascade transformer at High Voltages Laboratory (LAT-IEE/USP).

the High Voltages Laboratory (LAT) facilities of Electrotechnical and Energy Institute (IEE/USP) used for the essays from 10 kV_{RMS} up to 100 kV_{RMS}. The sensor interferometer connected to a high voltage cascade transformer, rated 1000 kV secondary, 150 kVA.

High-voltages at power frequency (60 Hz) up to 80 kV$_{RMS}$ were applied to the sensor interferometer. The experimental results from these tests and related linear regression curve are shown in Fig. 10.

From results shown in Fig. 10, it is observed that a linearity better them +/- 1,5% is obtained.

In addition, from other tests, it was possible to observe good temperature and time stabilities showed by the OVT, lower than 1000 ppm/°C in an observation period of half an hour.

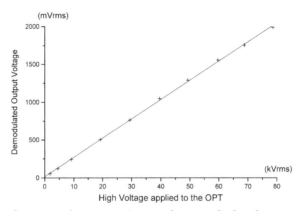

Fig. 10. Demodulated output voltage vs. ac input voltage applied to the sensor interferometer.

Such behaviors are due to the optical encoding technique as well as to the peculiar design of the electronic signal processing technique, which was specially developed for this application. The demodulation bandwidth of 1 kHz used makes possible to efficiently demodulate the firsts 16 harmonics of ac voltage applied to the sensor.

In Fig. 11 it is shown time-domain waveforms of demodulated output voltage and applied voltage (obtained from a reference divider) when about 70 kV$_{RMS}$ is applied to the OVT prototype.

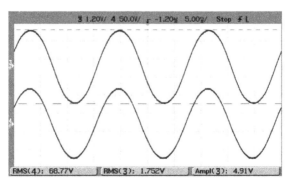

Fig. 11. Waveforms and RMS values of both the output demodulated voltage (channel 3) and ac high voltage applied to the sensor interferometer (channel 4).

In Fig. 12 frequency domain analyses are presented for both the output demodulated voltage and the ac high voltage applied to the sensor interferometer presented in Fig. 11.

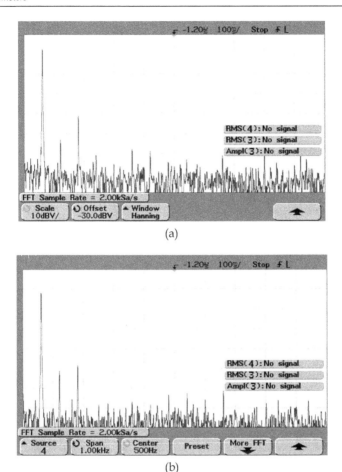

(a)

(b)

Fig. 12. Fast Fourier transforms, in the range from 0 kHz to 1 kHz: (a) of output
demodulated voltage and (b) of ac high voltage applied to sensor interferometer

Unfortunately, for 70 kV$_{RMS}$ applied to the sensor, the correspondent frequency domain
analysis observed in Fig.12(b) can't be extended for frequencies higher than 300 Hz due to
the bandwidth limitation of the reference divider used for measuring the amplitude of ac
high voltage applied to the sensor interferometer. In spite of that, a great likeness between
this spectrum and the one shown in Fig. 12(a) can be observed up to 300 Hz.

5.2 Normalization by an optical power reference sample technique

5.2.1 Principle of operation

To take advantage of WLI technique in a less expensive configuration capable to reduce the
dependency of the OVT output voltage, which carries out the measurement information,
$V_{ac}(t)$, on optical intensity fluctuations, an alternative encoding technique using an
unmodulated recover interferometer was proposed (Santos et al., 2003), as shown in Fig. 13.

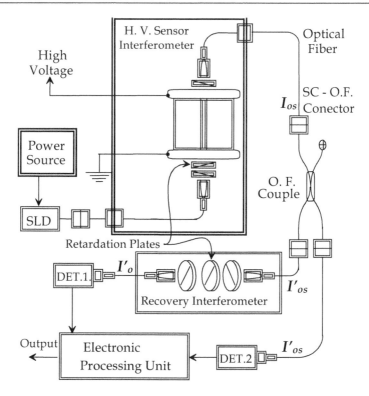

Fig. 13. OVT using WLI sensor system configuration with unmodulated recovery interferometer.

In this system, for recovering the interferometric signal a simple recovery interferometer, composed of a compensation plate inserted between two aligned polarizers, is placed in series with the optical fiber in the return path of sensing system by means of a directional coupler, which is an optical fiber beam splitter. From eq. (12), the optical intensity at the output of the recovery interferometer, I_o, is given by (Santos, 1996):

$$I_o = \frac{1}{4}\alpha_S\alpha_R T_1 T_2 I_i \left\{ 1 + \frac{1}{2}e^{-(\pi\Delta L/L_c)^2}\cos 2\pi\frac{\Delta L}{\lambda_0} \right\} \tag{38}$$

where I_i is the optical power at the input of the system (output of the SLD light source), α_S is the attenuation of light from the SLD to the input of sensor interferometer, λ_0 is the central wavelength of the light source and ΔL is the total OPD in the sensor system, given by: $\Delta L = \Delta L_S - \Delta L_c$.

The compensation plate used in the recovery interferometer is identical to the one used in the sensor interferometer. However, it is mounted in a special positioner, which can rotate around two axes to allow a fine tuning of the OPD introduced. Using this two axes rotation capability, it is possible to set the OPD of recovery interferometer in such way that: $\Delta L_{cr} = \Delta L_{cs} - \lambda_0/4$. In such condition, from eq. (38), I_o becomes:

$$I_o = \frac{1}{4}\alpha_S\alpha_R T_1 T_2 I_i \left\{ 1 + \frac{1}{2}e^{-\left[\frac{\lambda_0}{2}\left(\pi\frac{V}{V_\pi}-\frac{\pi}{2}\right)/L_c\right]^2} \cos\left(\pi\frac{V}{V_\pi}-\frac{\pi}{2}\right) \right\} \tag{39}$$

Since in practical applications L_c is much larger than λ_0, the exponential term in (39) remains close to the unity when $|V| < V_\pi$. Therefore, I_o can be approximated by:

$$I_o = I_{\bar{o}}\left(1 + \frac{1}{2}\sin\pi\frac{V}{V_\pi}\right) \tag{40}$$

where $I_{\bar{o}} = \frac{1}{4}\alpha_S\alpha_R T_1 T_2 I_i = \frac{1}{4}\alpha_t I_i$, is the average of output light intensity.

The response curve for this kind of WLI sensor system is similar to the response curve of a simple polarimetric Pockels modulator and, in this case, also there is a dependency of output light intensity, I_o, on the total attenuation of system. To take advantage of WLI technique and eliminate such dependency, an electronic signal processing scheme was developed, using as reference a sample of light intensity at output of sensor interferometer, I_{os}, which is given by:

$$I_{os} = \frac{1}{2}\alpha_s T_1 T_2 I_i \left\{ 1 + e^{-(\pi\Delta L_s/L_c)^2} \cos 2\pi\frac{\Delta L_s}{\lambda_0} \right\} \tag{41}$$

Since $\Delta L_s >> L_c$, the exponential term in eq. (40) approaches to zero. Therefore, I_{os} reduces to:

$$I_{os} = \frac{1}{2}\alpha_s T_1 T_2 I_i = \frac{2}{\alpha_R}I_{\bar{o}} \tag{42}$$

As can be seem in Fig. 13, placing a directional coupler between the output of sensor interferometer and the input of recover interferometer it is possible to obtain the signal $I'_{os}(t)$, which is detected by the photodetector (DET. 2) and used to compensate the signal obtained in the output of recover interferometer, I'_o, for any optical power variation. Using a 3 dB directional coupler, the incoming optical intensity is divided in equal parts between the two outputs. Therefore, the optical power I'_{os}, which reaches the photodetector (DET. 2), is:

$$I'_{os} = \alpha_{ex} I_{os}/2 \tag{43}$$

where α_{ex} is the excess loss of directional coupler and I_{os} is the optical intensity at output of sensor interferometer.

The optical intensity at output of recovery interferometer, $I'_o(t)$, is given by:

$$I'_o(t) = \frac{1}{2}\alpha_r I'_{os}\left(1 + \frac{1}{2}\sin\pi\frac{V_{ac}(t)}{V_\pi}\right) \tag{44}$$

where α_r is the total attenuation of recovery interferometer, I'_{os} is the optical power at output of optical fiber coupler and V_π is the half-wave voltage of sensor interferometer.

A blocks diagram of electronic signal processing unit required by such system is given in Fig. 14.

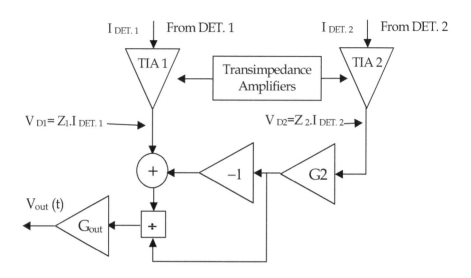

Fig. 14. Blocks diagram of electronic signal processing unit.

Applying the signals I'_{os} and $I'_o(t)$ to the electronic signal processing unit proposed in Fig. 14, if the transimpedance gains Z_1 and Z_2 are equals and the gain $G_2 = \alpha_r/2$, the output signal, $V_{out}(t)$, can be linearized (when $V_{ac}(t) \ll V_\pi$) and approximated by the following expression:

$$V_{out}(t) \cong \frac{G_{out}}{2} \cdot \pi \cdot \frac{V_{ac}(t)}{V_\pi} \tag{45}$$

where G_{out} is the gain of last amplifier stage.

From eq. (45) it is clear that $V_{out}(t)$ is independent on the optical intensity at optical fiber link, as desired.

5.2.2 Experimental results

To verify the V_{out} dependency on optical power I_{os}, an experimental setup implementing the system described in Fig. 13 was built using prototypes of high voltage Pockels cell and recovery interferometer previously developed (Santos et al., 2000, 2003), a 3 dB directional coupler (commercially available), a SLD operating in a wavelength of 1.3 μm as broadband light source and an electronic signal processing unit developed according to the blocks diagram shown in Fig. 14. To introduce variations in I_{os}, an adjustable optical attenuator was inserted between the output of sensor interferometer and the input of directional coupler.

Experiments were conducted changing the attenuation imposed by the adjustable attenuator in a range of 0 to –8.66 dB. During the variation of the attenuation the voltage applied to the OVT was maintained constant in a reference value of 4.5 kV_{pp}. In Table 2, bellow, the attenuation values introduced by the attenuator, the RMS values of measured signals in the electronic processing unit output and the relative variations of correspondent V_{out} value, named ΔV_{out}, are shown.

Attenuation (dB)	V_{out} (mV$_{RMS}$)	ΔV_{out} (%)
0	131,9	0
-2,5	131,0	-0,7
-3,46	131,0	-0,7
-4,75	131,1	-0,6
-5,54	132,0	0,1
-6,51	132,0	0,1
-7,49	133,2	1,0
-8,66	133,0	0,8

Table 2. Attenuation values introduced by the attenuator, RMS values of measured signals in the electronic processing unit output and relative variations of the V_{out} values (ΔV_{out}).

From values presented in table 2, it is shown that, for a variation of about -9 dB in average optical power carried by the optical fiber link, less than 1% of variation is observed in output of WLI sensor system.

6. Conclusions

Both alternatives of WLI based OVTs presented have adequate performances for measuring ac high voltages while keeping low sensibility to fluctuations on average light intensity in the optical fiber link.

The advantages exhibited by the first approach, compared to the second one, are the better characteristics in terms of resolution and thermal and temporal stabilities. However, the cost and complexity of that kind of solution are higher than the second one. Such characteristics make first solution suitable for applications where high performance is demanded.

The main advantages of the second approach are the lower cost and smaller complexity of the entire system. Therefore, it is recommended for applications demanding moderate performances and lower prices.

New optical encoding techniques and electronic signal processing techniques applied to WLI sensor systems are still in development, and it is expected that many improvements arise in this field in a near future.

7. Acknowledgment

The authors would like to thank to the University of São Paulo – USP - and to São Paulo Research Foundation - FAPESP - for the financial support to this work.

8. References

Almeida, J. C. J. de & Filho, O. V. A. (2003). *Demodulador de desvio de fase óptico não recíproco num sensor óptico interferométrico, via detecção dos valores dos picos da corrente detectada no fotodetector acoplado à saída do interferômetro óptico*, patent required to the Instituto Nacional da Propriedade Industrial - INPI, according to the registry number PI0303.688-0, Brazil

Almeida, J. C. J. de, (2001). *Nova Técnica de Processamento de Sinal no Domínio do Tempo de Giroscópios Interferométricos de Sagnac a Fibra Óptica*, Ph.D. Thesis, State University of Campinas, Brazil

Bull, J. D., Jaeger, N. A. F., & Rahmatian, F. (2005). A new hybrid current sensor for high-voltage applications. IEEE Transactions on Power Delivery, Vol. 20, No. 1, (Jan 2005), pp. 32-38, ISSN 0885-8977.

Burns, W. K. (1994). *Optical Fiber Rotation Sensing*, (Ed. 1), Academic Press, Inc., ISBN 0-1214.6075-4, Boston, USA

Chen, S.; Palmer, A. W.; Grattan, K. T. V.; & Meggit, B. T. (1992). Digital signal-processing techniques for electronically scanned optical-fiber white-light interferometry, *Applied Optics*, Vol. 31, No. 28, (October 1992), pp. 6003-6010, ISSN 1559-128X

Chen, X.; Ye, W.; Zhang, H. & Liu, T. (2010). Spectral domain demodulation of fibre optics position and displacement sensors by Fourier-transform spectral interferogram. *Proceedings of 9th International Conference on Optical Communications and Networks (ICOCN 2010)*, ISBN 978-1-84919-314-6, Nanjing, China, October 2010.

Crotti , G.; Sardi, A.; Kuljaca, N.; Mazza, P.; De Donà, G.; Brand, U.; Giraud, M.; A. Andersson & Weiss, S. (2006). "On-site live verification of HV instrument transformer accuracy", CIGRE General Session 41, Aug. 27 - Sep. 1, 2006, paper A3-204

Franco, M. A. R.; Passaro, A.; Sircilli, F. & Cardoso, J. R. (1999). Finite element analysis of anisotropic optical waveguide with arbitrary index profile. *IEEE Transactions on Magnetics*, Vol.35, No. 3, (May 1999), pp. 1546-1549, ISSN 0018-9464

Françon, M. (1967). *Optical Interferometry*, Academic Press, ISBN 0-1226.6350-0, New York, USA

Kim, J.H. (2008). An All Fiber White Light Interferometric Absolute Temperature Measurement System. Sensors, 8, 6825-6845. ISSN 1424-8220

Lequine, M.; Lecot, C.; Giovannini, H.; & Huard, S. (1990). A dual wavelegth passive homodyne detection unit for fiber coupled white light interferometers, In: *Fiber-Optic Sensors IV*, Vol. 1267, Editor: Ralf T. Kersten, pp. 288-293, Proceedings of SPIE, ISBN 9780819403148

Liao, H. & Yang, Y. (2010). A Linnik Scanning White-Light Interferometry System Using a MEMS Digital-to-Analog Converter, Proc. Eurosensors XXIV, September 5-8, 2010, Linz, Austria Volume 5, 2010, Pages 758-761, ISSN: 1877-7058.

Mariscotti, A. (2009). A Rogowski winding with high voltage immunity, In: EUROCON 2009 pp. 1129-1133, ISBN 978-1-4244-3860-0, St. Petersburg, RU, May 18-23 2009.

Murtaza, G. (1991). *Dual Wavelength referenced intensity modulated optical fibre sensor system*, PhD Thesys, Manchester Polytechnic, CNAA

Ning, Y. N.; Grattan, K. T. V.; & Palmer, A.W. (1995). In: *Applications of Photonic Technology*, Editors: Lampropoulos, G. A.; Chrostowsky, J. & Measures, R. M., pp. 339-342, Plenum Press, ISBN 0-3064.5011-9

Rahmatian, F. (2010). High-voltage current and voltage sensors for a smarter transmission grid and their use in live-line testing and calibration, In: *IEEE Power and Energy Society General Meeting*, Minneapolis, MN, ISSN: 1944-9925, E-ISBN: 978-1-4244-8357-0, Print ISBN: 978-1-4244-6549-1, July 2010

Rubini Jr., J.; A. Passaro, Abe, N. M. & Santos, J. C. (2004) Analysis of the electric field distribution in an electrooptic sensor for pulsed high-voltage measurements, *Proceedings of Fifth IEE International Conference on Computation in Electromagnetics – CEM2004*, pp. 71-72, ISBN 0-8634.1400-1, Stratford-upon-Avon, UK, April 19-22, 2004

Santos, J. C. (2009). Contribuições para o desenvolvimento de transformadores de potencial a fibras ópticas (TPs Ópticos) aplicáveis em sistemas elétricos de potência, L.D. Thesis, University of São Paulo, Brazil.

Santos, J. C.; Côrtes, A. L. & Hidaka, K. (1999). A New Electro-optical Method for recovering White Light Interferometric Signals, *Proceedings of 1999 International Microwave And Optoeletronics Conference - IMOC'99*, ISBN 0-7803-5807-4, Rio de Janeiro, Brazil, August 1999

Santos, J. C.; Côrtes, A. L.; Hidaka, K. & Silva, L. P. C. (2003). Improved optical sensor for high voltage measurement using white light interferometry, *Proceedings of SBMO/IEEE MTT-S IMOC 2003*, ISBN 0-7803-7824-5, Fóz do Iguacú, Paraná, Brazil, September 2003

Santos, J.C.; Hidaka, K. & Côrtes, A. L. (2002). Optical High Voltage Measurement Transformer Using White Light Interferometry, *Proceedings of 2002 IEEE/PES T&D Latin America Conference*, in CD-ROM, São Paulo, Brazil, March 2002

Santos, J.C.; Taplamacioglu, M. C. & Hidaka, K. (2000) Pockels high-voltage measurement system. *IEEE Transactions on Power Delivery*, Vol.15, No.1, (January 2000), pp. 8–13, ISSN 0885-8977

Tselikov, A.; Arruda, J. U. & Blake, J. (1998). Zero-Crossing Demodulation for Open Loop Sagnac Interferometers, *IEEE Journal of Lightwave Technology*, Vol. 16, No. 9, (September 1998), pp. 1613-1619, ISSN 0733-8724

Velosa, J.; Gouveia, C.; Frazao, O.; Jorge, P. & Baptista, J. (2011). Digital Control of a White Light System for Optical Fiber Interferometers. *IEEE Sensors Journal*, Vol. PP, No. 99, (April 2011), pp. 1 – 1, ISSN: 1530-437X

Yariv, A. & Yeh, P. (1984). *Optical waves in cristals*. (Ed. 1), Wiley-Interscience, ISBN 0-4714.3081-1, New York, USA

Yu, B.; A. Wang, Gary Pickrell & Xu, J. (2005) "Tunable-optical-filter-based white-light interferometry for sensing," Opt. Lett. 30, 1452-1454 (2005), Optics Letters, Vol. 30, Issue 12, pp. 1452-1454, ISSN: 0146-9592

Permissions

The contributors of this book come from diverse backgrounds, making this book a truly international effort. This book will bring forth new frontiers with its revolutionizing research information and detailed analysis of the nascent developments around the world.

We would like to thank Dr Moh. Yasin, Prof. Sulaiman W. Harun and Dr Hamzah Arof, for lending their expertise to make the book truly unique. They have played a crucial role in the development of this book. Without their invaluable contribution this book wouldn't have been possible. They have made vital efforts to compile up to date information on the varied aspects of this subject to make this book a valuable addition to the collection of many professionals and students.

This book was conceptualized with the vision of imparting up-to-date information and advanced data in this field. To ensure the same, a matchless editorial board was set up. Every individual on the board went through rigorous rounds of assessment to prove their worth. After which they invested a large part of their time researching and compiling the most relevant data for our readers. Conferences and sessions were held from time to time between the editorial board and the contributing authors to present the data in the most comprehensible form. The editorial team has worked tirelessly to provide valuable and valid information to help people across the globe.

Every chapter published in this book has been scrutinized by our experts. Their significance has been extensively debated. The topics covered herein carry significant findings which will fuel the growth of the discipline. They may even be implemented as practical applications or may be referred to as a beginning point for another development. Chapters in this book were first published by InTech; hereby published with permission under the Creative Commons Attribution License or equivalent.

The editorial board has been involved in producing this book since its inception. They have spent rigorous hours researching and exploring the diverse topics which have resulted in the successful publishing of this book. They have passed on their knowledge of decades through this book. To expedite this challenging task, the publisher supported the team at every step. A small team of assistant editors was also appointed to further simplify the editing procedure and attain best results for the readers.

Our editorial team has been hand-picked from every corner of the world. Their multi-ethnicity adds dynamic inputs to the discussions which result in innovative outcomes. These outcomes are then further discussed with the researchers and contributors who give their valuable feedback and opinion regarding the same. The feedback is then collaborated with the researches and they are edited in a comprehensive manner to aid the understanding of the subject.

Apart from the editorial board, the designing team has also invested a significant amount of their time in understanding the subject and creating the most relevant covers. They scrutinized every image to scout for the most suitable representation of the subject and create an appropriate cover for the book.

The publishing team has been involved in this book since its early stages. They were actively engaged in every process, be it collecting the data, connecting with the contributors or procuring relevant information. The team has been an ardent support to the editorial, designing and production team. Their endless efforts to recruit the best for this project, has resulted in the accomplishment of this book. They are a veteran in the field of academics and their pool of knowledge is as vast as their experience in printing. Their expertise and guidance has proved useful at every step. Their uncompromising quality standards have made this book an exceptional effort. Their encouragement from time to time has been an inspiration for everyone.

The publisher and the editorial board hope that this book will prove to be a valuable piece of knowledge for researchers, students, practitioners and scholars across the globe.

List of Contributors

Julián M. Estudillo-Ayala, Ruth I. Mata-Chávez and Roberto Rojas-Laguna
Universidad de Guanajuato, México

Juan C. Hernández-García
Centro de Investigaciones en Óptica A.C., México

Catarina Silva and João M. P. Coelho
Departamento de Física, Faculdade de Ciências da Universidade de Lisboa, Portugal

Paulo Caldas
Departamento de Física, Faculdade de Ciências da Universidade do Porto, Portugal
Unidade de Optoelectrónica e Sistemas Electrónicos, INESC Porto, Portugal
Escola Superior de Tecnologia e Gestão de Viana do Castelo, Portugal

Pedro Jorge
Unidade de Optoelectrónica e Sistemas Electrónicos, INESC Porto, Portugal

H. Z. Yang and H. Ahmad
Photonic Research Center, University of Malaya, Kuala Lumpur, Malaysia

S. W. Harun
Photonic Research Center, University of Malaya, Kuala Lumpur, Malaysia
Department of Electrical Engineering, Faculty of Engineering, University of Malaya, Kuala Lumpur, Malaysia

M. Yasin
Photonic Research Center, University of Malaya, Kuala Lumpur, Malaysia
Department of Physics, Faculty of Science and Technology, Airlangga University, Surabaya, Indonesia

Marta S. Ferreira, Ricardo M. Silva and Orlando Frazão
INESC Porto, Porto, Portugal

Kevin S. C. Kuang
National University of Singapore, Department of Civil and Environmental Engineering, Singapore

Dongsheng Li, Liang Ren and Hongnan Li
Dalian University of Technology, China

Luigi Rovati and Luca Ferrari
Department of Information Engineering, University of Modena and Reggio Emilia, Italy

Paola Fabbri and Francesco Pilati
Department of Materials and Environmental Engineering, University of Modena and Reggio Emilia, Italy

Masataka Iinuma, Yasuyuki Ushio, Akio Kuroda and Yutaka Kadoya
Graduate School of Advanced Sciences of Matter, Hiroshima University, Japan

Mario Francisco Jesús Cepeda Rubio, Arturo Vera Hernández and Lorenzo Leija Salas
Centro de Investigación y de Estudios Avanzados del Instituto Politécnico Nacional, Department of Electrical Engineering/Bioelectronics, Mexico

Josemir C. Santos and Luiz P. C. Silva
University of São Paulo, Brazil

José C. J. Almeida
Optsensys, Brazil

Printed in the USA
CPSIA information can be obtained
at www.ICGtesting.com
JSHW011421221024
72173JS00004B/624

9 781632 381941